Released from
Samford University Library

The Adventure Playground of Mechanisms and Novel Reactions

Rolf Huisgen

PROFILES, PATHWAYS, AND DREAMS
Autobiographies of Eminent Chemists

Jeffrey I. Seeman, Series Editor

D0209816

American Chemical Society, Washington, DC 1994

Samford University Library

Library of Congress Cataloging-in-Publication Data

Rolf Huisgen
The adventure playground of mechanisms and novel reactions / Rolf
Huisgen

p.   cm.—(Profiles, pathways, and dreams, ISSN 1047–8329)
Includes bibliographical references and index.
ISBN 0–8412–1832–3

1. Huisgen, Rolf. 2. Chemists—Germany—Biography.
3. Chemistry, Organic—History—20th century.

I. Title. II. Series.

QD22.H86A3   1994
540'.92—dc20   94–13957
[B]   CIP
Jeffrey I. Seeman, Series Editor

The paper used in this publication meets the minimum requirements of American
National Standard for Information Sciences—Permanence of Paper for Printed Library
Materials, ANSI Z39.48–1984.

Copyright © 1994

American Chemical Society

All Rights Reserved. The copyright owner consents that reprographic copies may be
made for personal or internal use or for the personal or internal use of specific clients.
This consent is given on the condition, however, that the copier pay the stated per-
copy fee through the Copyright Clearance Center, Inc., 27 Congress Street, Salem, MA
01970, for copying beyond that permitted by Sections 107 or 108 of the U.S. Copyright
Law. This consent does not extend to copying or transmission by any means—graphic
or electronic—for any other purpose, such as for general distribution, for advertising or
promotional purposes, for creating a new collective work, for resale, or for information
storage and retrieval systems. The copying fee is $0.75 per page. Please report your
copying to the Copyright Clearance Center with this code: 1047–8329/93/$00.00+.75.

The citation of trade names and/or names of manufacturers in this publication is not to
be construed as an endorsement or as approval by ACS of the commercial products or
services referenced herein; nor should the mere reference herein to any drawing,
specification, chemical process, or other data be regarded as a license or as a
conveyance of any right or permission to the holder, reader, or any other person or
corporation, to manufacture, reproduce, use, or sell any patented invention or
copyrighted work that may in any way be related thereto. Registered names,
trademarks, etc., used in this publication, even without specific indication thereof, are
not to be considered unprotected by law.

PRINTED IN THE UNITED STATES OF AMERICA

1994 Advisory Board

QD
22
. H 86
A3
1994

Robert J. Alaimo
Procter & Gamble Pharmaceuticals

Mark Arnold
University of Iowa

David Baker
University of Tennessee

Arindam Bose
Pfizer Central Research

Robert F. Brady, Jr.
Naval Research Laboratory

Margaret A. Cavanaugh
National Science Foundation

Arthur B. Ellis
University of Wisconsin at Madison

Dennis W. Hess
Lehigh University

Hiroshi Ito
IBM Almaden Research Center

Madeleine M. Joullie
University of Pennsylvania

Lawrence P. Klemann
Nabisco Foods Group

Gretchen S. Kohl
Dow-Corning Corporation

Bonnie Lawlor
Institute for Scientific Information

Douglas R. Lloyd
The University of Texas at Austin

Cynthia A. Maryanoff
R. W. Johnson Pharmaceutical
Research Institute

Julius J. Menn
Western Cotton Research Laboratory,
U.S. Department of Agriculture

Roger A. Minear
University of Illinois
at Urbana–Champaign

Vincent Pecoraro
University of Michigan

Marshall Phillips
Delmont Laboratories

George W. Roberts
North Carolina State University

A. Truman Schwartz
Macalaster College

John R. Shapley
University of Illinois
at Urbana–Champaign

L. Somasundaram
DuPont

Michael D. Taylor
Parke-Davis Pharmaceutical Research

Peter Willett
University of Sheffield (England)

Foreword

In 1986, the ACS Books Department accepted for publication a collection of autobiographies of organic chemists, to be published in a single volume. However, the authors were much more prolific than the project's editor, Jeffrey I. Seeman, had anticipated, and under his guidance and encouragement, the project took on a life of its own. The original volume evolved into 22 volumes, and the first volume of Profiles, Pathways, and Dreams: Autobiographies of Eminent Chemists was published in 1990. Unlike the original volume, the series was structured to include chemical scientists in all specialties, not just organic chemistry. Our hope is that those who know the authors will be confirmed in their admiration for them, and that those who do not know them will find these eminent scientists a source of inspiration and encouragement, not only in any scientific endeavors, but also in life.

M. Joan Comstock
Head, Books Department
American Chemical Society

Contributors

We thank the following corporations and Herchel Smith for their generous financial support of the series Profiles, Pathways, and Dreams.

Akzo nv

Bachem Inc.

E. I. du Pont de Nemours
and Company

Duphar B.V.

Eisai Co., Ltd.

Fujisawa Pharmaceutical Co., Ltd.

Hoechst Celanese Corporation

Imperial Chemical Industries PLC

Kao Corporation

Mitsui Petrochemical Industries,
Ltd.

The NutraSweet Company

Organon International B.V.

Pergamon Press PLC

Pfizer Inc.

Philip Morris

Quest International

Sandoz Pharmaceuticals
Corporation

Sankyo Company, Ltd.

Schering–Plough Corporation

Shionogi Research Laboratories,
Shionogi & Co., Ltd.

Herchel Smith

Suntory Institute for Bioorganic
Research

Takasago International
Corporation

Takeda Chemical Industries, Ltd.

Unilever Research U.S., Inc.

Profiles, Pathways, and Dreams

Titles in This Series

About the Editor

JEFFREY I. SEEMAN received his B.S. with high honors in 1967 from the Stevens Institute of Technology in Hoboken, New Jersey, and his Ph.D. in organic chemistry in 1971 from the University of California, Berkeley. Following a two-year staff fellowship at the Laboratory of Chemical Physics of the National Institutes of Health in Bethesda, Maryland, he joined the Philip Morris Research Center in Richmond, Virginia. In 1983–1984, he enjoyed a sabbatical year at the Dyson Perrins Laboratory in Oxford, England, and claims to have visited more than 90% of the castles in England, Wales, and Scotland.

Seeman's 90 published papers include research and patents in the areas of photochemistry, nicotine and tobacco alkaloid chemistry and synthesis, conformational analysis, pyrolysis chemistry, organotransition metal chemistry, the use of cyclodextrins for chiral recognition, and structure–activity relationships in olfaction. He was a plenary lecturer at the Eighth IUPAC Conference on Physical Organic Chemistry held in Tokyo in 1986 and has been an invited lecturer at numerous scientific meetings and universities. Currently, Seeman serves on the Petroleum Research Fund Advisory Board. He continues to count Nero Wolfe and Archie Goodwin among his best friends.

Contents

Photographs

Preface

"HOW DID YOU GET THE IDEA—and the good fortune—to convince 22 world-famous chemists to write their autobiographies?" This question has been asked of me, in these or similar words, frequently over the past several years. I hope to explain in this preface how the project came about, how the contributors were chosen, what the editorial ground rules were, what was the editorial context in which these scientists wrote their stories, and the answers to related issues. Furthermore, several authors specifically requested that the project's boundary conditions be known.

As I was preparing an article[1] for *Chemical Reviews* on the Curtin–Hammett principle, I became interested in the people who did the work and the human side of the scientific developments. I am a chemist, and I also have a deep appreciation of history, especially in the sense of individual accomplishments. Readers' responses to the historical section of that review encouraged me to take an active interest in the history of chemistry. The concept for Profiles, Pathways, and Dreams resulted from that interest.

My goal for Profiles was to document the development of modern organic chemistry by having individual chemists discuss their roles in this development. Authors were not chosen to represent my choice of the world's "best" organic chemists, as one might choose the "baseball all-star team of the century". Such an attempt would be foolish: Even the selection committees for the Nobel prizes do not make their decisions on such a premise.

xvii

The selection criteria were numerous. Each individual had to have made seminal contributions to organic chemistry over a multidecade career. (The average age of the authors is over 70!) Profiles would represent scientists born and professionally productive in different countries. (Chemistry in 13 countries is detailed.) Taken together, these individuals were to have conducted research in nearly all subspecialties of organic chemistry. Invitations to contribute were based on solicited advice and on recommendations of chemists from five continents, including nearly all of the contributors. The final assemblage was selected entirely and exclusively by me. Not all who were invited chose to participate, and not all who should have been invited could be asked.

A very detailed four-page document was sent to the contributors, in which they were informed that the objectives of the series were

1. to delineate the overall scientific development of organic chemistry during the past 30–40 years, a period during which this field has dramatically changed and matured;

2. to describe the development of specific areas of organic chemistry; to highlight the crucial discoveries and to examine the impact they have had on the continuing development in the field;

3. to focus attention on the research of some of the seminal contributors to organic chemistry; to indicate how their research programs progressed over a 20–40-year period; and

4. to provide a documented source for individuals interested in the hows and whys of the development of modern organic chemistry.

One noted scientist explained his refusal to contribute a volume by saying, in part, that "it is extraordinarily difficult to write in good taste about oneself. Only if one can manage a humorous and light touch does it come off well. Naturally, I would like to place my work in what I consider its true scientific perspective, but . . ."

Each autobiography reflects the author's science, his lifestyle, and the style of his research. Naturally, the volumes are not uniform, although each author attempted to follow the guidelines. "To write in good taste" was not an objective of the series. On the contrary, the authors were specifically requested not to write a review article of their field, but to detail their own research accomplishments. To the extent that this instruction was followed and the result is not "in good taste", then these are criticisms that I, as editor, must bear, not the writer.

As in any project, I have a few regrets. It is truly sad that Egbert Havinga and Herman Mark, who each wrote a volume, and David Ginsburg, who translated another, died during the course of this project. There have been many rewards, some of which are documented in my personal account of this project, entitled "Extracting the Essence: Adventures of an Editor" published in *CHEMTECH*.[2]

Acknowledgments

I join the entire scientific community in offering each author unbounded thanks. I thank their families and their secretaries for their contributions. Furthermore, I thank numerous chemists for reading and reviewing the autobiographies, for lending photographs, for sharing information, and for providing each of the authors and me the encouragement to proceed in a project that was far more costly in time and energy than any of us had anticipated.

I thank my employer, Philip Morris USA, and J. Charles, R. N. Ferguson, K. Houghton, and W. F. Kuhn, for without their support Profiles, Pathways, and Dreams could not have been. I thank ACS Books, and in particular, Robin Giroux (production manager), Janet Dodd (senior editor), Joan Comstock (department head), and their staff for their hard work, dedication, and support. Each reader no doubt joins me in thanking 24 corporations and Herchel Smith for financial support for the project.

I thank my children, Jonathan and Brooke, for their patience and understanding; remarkably, I have been working on Profiles for more than half of their lives—probably the only half that they can remember! Finally, I again thank all those mentioned and especially my family, friends, colleagues, and the 22 authors for allowing me to share this experience with them.

JEFFREY I. SEEMAN
Philip Morris Research Center
Richmond, VA 23234

April 7, 1992

[1] Seeman, J. I. *Chem. Rev.* **1983**, *83*, 83–134.
[2] Seeman, J. I. *CHEMTECH* **1990**, *20*(2), 86–90.

It is an indisputable requirement in organic chemistry that we refrain from stripping the phenomena, not yet clarified, of the charm that lies precisely in their obscurity.

Justus von Liebig, 1855

Editor's Note

ROLF HUISGEN FOLLOWED IN THE FOOTSTEPS of his teacher, mentor, and friend, Heinrich Wieland, when he accepted the prestigious chair of organic chemistry (three Nobel Prize winners in a row!) at the University of Munich in 1952. At the age of 32, Huisgen was nearly a decade younger than the youngest of his predecessors![*] Certainly, some must have wondered if the young chemist could continue the scientific legacy.

The Munich laboratory had been totally destroyed during World War II (*see* page 6), and attempts to win Richard Kuhn or Clemens Schöpf as Wieland's successor had failed. Rolf Huisgen explained, "I received the chair not as a reward for scientific merits, but rather as a trust for the future." He commented on the philosophy that led him to accept the challenge: "Are not most nominations for professorships lottery tickets? The creative spirit may vanish, or—in the case of a young scientist—never evolve. . . . I have persistence in pursuing a goal." His wife Trudl summed him up well: "Intellectual curiosity combined with strong ambition."

With his research, Rolf Huisgen made the Munich laboratory (once a stronghold of natural product chemistry) one of the foremost German schools in physical organic chemistry and reaction mechanisms. He maintained it as, according to a leading European academician, " the most important chemical institution in Germany." As acknowledged in

[*]H. Wieland, 48; R. Willstätter, 47; A. von Baeyer, 40; and J. von Liebig, 49

his award citation for the Otto Hahn Prize of Chemistry and Physics (German Chemical Society, 1979):

> *The rich scientific harvest, brought in by more than 200 gradu-ate students and postdoctoral fellows from all parts of the world, was set forth in nearly 500 publications, which made Rolf Huis-gen the most cited German chemist. His work intrigues through elegance of argumentation, succinctness of style, and through the reliability of the experimental results.*

Electronic theory was not very popular in German chemistry at the end of World War II, despite the efforts of Hans Meerwein, Fritz Arndt, Bernd Eistert, and Walter Hückel. According to Huisgen, "Formulas with many dots for electrons were labeled *fly feces formulae* by some aged pundits." Huisgen helped change that perception and guide the growth of modern mechanistic theory in Germany. "Rolf is very strong-willed and self-disciplined," his wife, Trudl, told me. "He did not like administrative work and exams, but did both with a high sense of duty. Rolf is a loyal person." Teaching was more to his liking. To quote from the Otto Hahn award: "His lectures, internationally appre-ciated, impressed the audience by transparence and polished diction."

Manfred Schlosser, a professor of organic chemistry at the Univer-sity of Lausanne, described Huisgen as "rigorous intellectually and experimentally. He is so full of discipline, and he is a very warm per-son. Few people see this warmth." Huisgen describes himself as "shy by nature" and does indeed consider himself somewhat formal. But this is a man who can promote and share warm friendships and laughter. Once, when teased about always wearing a tie, he quipped, "I use a rubber one in the shower!"

After several years of formal though friendly interaction, our rela-tionship evolved to a friendly give-and-take of jokes and humor, and he often joined me in hearty laughter that made us both feel good. I recall one tongue-in-cheek chiding he gave me when I had gotten off schedule with his galleys: "Your delay in reading the galleys did not remain hidden to me," he wrote. "To publish the volume still in this millennium, my suggestions would be. . . ." Generous with compli-ments, he assured me that I should be able to "handle the next dozen authors with ease."

Schlosser recalls going skiing with Huisgen many years ago. "We had so much fun! He wore a fantastic outfit, family property since the last century. We moved to the edge of this mountain. I looked down; it was a steep slope—almost a free fall. Huisgen simply went over! I took a detour and found him ten minutes later. 'How could you do this?' I asked. 'It is a little steep,' he responded, 'but sooner or later, it goes up again and you can stop.'" Another facet of his personality is his thoroughness. For example, in preparing his manuscript he compiled over 600 references that he felt were necessary to achieving a full and accurate account of his life's work. What might be overwhelming for many is simply the most direct route to meet his standards, analogous to his taking that direct course downhill because it is there.

To prepare for this Editor's Note, I asked some soul-searching questions of my author and friend: What do you feel are your most important qualities? His response was in classical Greek:

$$\text{Γνῶθι σεαυτον}$$

(know thyself), inscribed on the Apollo Temple at Delphi, is well known.

Then he continued

- Not to be content with analogies, but to ask for reasons within the limited realm of cognition.

- An intellectual pleasure in connecting seemingly unrelated phenomena that sometimes leads to new principles.

- A moderately good memory, one of the prerequisites of survival in chemistry.

- A love of detail, as that is the touchstone for the usefulness of principles and ideas.

Huisgen's gifts of scholarship, intellectual passion, and infectious inspiration are indeed best revealed by his own words.

I then asked what career would he have chosen had he not become a scientist. His immediate response was "an art historian or an archeologist." Art is his passion. A museum or galley tour with him encapsulates art, history, and philosophy. I look forward to his Christmas cards because he reproduces art from his own collection, accompanied by his own profound commentary. Huisgen admits "to the caprice of stockpiling aphorisms by classic and modern authors. We should allow them to enter the chemical literature; they often express a conclusion most concisely." Some years ago Klee's "Rope-dancer" on such a card inspired his comment that it was a "fitting symbol for human life and aspirations; keeping the balance requires incessant corrections. It might even be a metaphor for fate and fortune of nations, the protection of peace among nations being a tightrope act."

"What have been your most important lessons in life?" was my final question.

"To avoid superlatives. Life is painted in many shades of gray. Not to trust slogans and simplistic solutions, neither in science nor in social life, nor in politics. To rid myself of prejudices—a Herculean labor—because education, schools, textbooks, research publications, and contacts with family and friends build up prejudices."

JEFFREY I. SEEMAN
Philip Morris Research Center
Richmond, VA 23234

April 15, 1994

As editor, I take the liberty of reproducing, with Rolf Huisgen's kind permission, a pen and ink sketch by his brother, Klaus. Rolf felt that the illustration which appears on page 207 was sufficient, but given the closeness of the brothers and their artistic and philosophic connections—and the subject of this sketch—I share "Refugees" (France, 1940) with the reader.

The Adventure Playground

of Mechanisms

and Novel Reactions

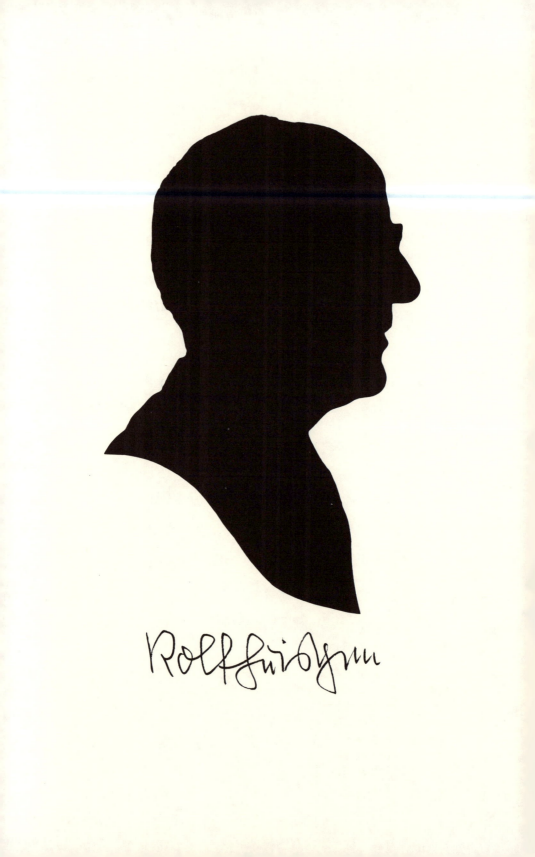

The Setting

My First Encounter with Chemistry

Paul Walden's *Geschichte der Organischen Chemie seit 1880* (History of Organic Chemistry since 1880)[1] gives a marvelous account of the stunning post-Liebig evolution. However, the book did not receive the worldwide attention it deserved because it was published during World War II. While I was an associate professor at the University of Tübingen (1949–1952), I heard Paul Walden lecture. Walden, by then nearly a nonagenarian, had found refuge in Gammertingen in the Swabian mountains after the horrors of the war. His grand historical discourses offered a bird's-eye view of the development of organic chemistry, but did not reach quite as far as 1950. A diminished appreciation of very recent progress is possibly an inevitable consequence of old age. As many deaf or nearly deaf people do, Walden had lost the sense of the volume of his voice. During the luncheon after his lectures, it was not only his colleagues who enjoyed the privilege of listening to his witty monologues.

A substantial part of the scientific and industrial evolution of chemistry before 1940 was shaped in Germany. The names of Nobel Laureates R. Willstätter (1915), H. Wieland (1927), A. Windaus (1929), H. Fischer (1930), R. Kuhn (1938), and A. Butenandt (1939) evoke the heroic phase when the structures of steroids, plant pigments, vitamins, and hormones were elucidated. Organic chemistry then was to a large degree a descriptive science governed by heuristic principles; the young disciples of the art were soon confronted with a maze.

My father, Edmund Huisgen, a surgeon, fostered my interest in the natural sciences. Thus, I began to study mathematics and physics at

3

the University of Bonn in 1939. But then, enthralled by Paul Pfeiffer's brilliant introductory course in organic chemistry, I enrolled in chemistry at the University of Munich in my second semester. Munich had one of the four universities still operating at the beginning of the war. The work on my thesis (Ph.D. 1943), directed by Heinrich Wieland,[2] concerned the chemistry of vomicine, a strychnine alkaloid.[3]

Heinrich Wieland (1877–1957), one of the master architects of chemical sciences, directed the chemical laboratory of the University of Munich from 1926 to 1952. Bronze by Emil Krieger, cast in 1952.

Heinrich Wieland was respected by his contemporaries as one of the last universalists of chemistry.[4a] His research interests ranged from classical organic topics, like the oligomerization of fulminic acid, to highlights of natural product chemistry, such as the structure of the bile acids, the amanita toxin, or the *Lobelia* alkaloids. The young generation hardly remembers that Wieland was actually one of the fathers of biochemistry. His 53 papers, *Zum Mechanisms der Oxidationsvorgänge* (On the Mechanisms of Oxidation Processes), concern fundamentals like hydrogen transfer in biological oxidation and the key role of acetic acid in metabolism.[4b] On the inorganic side, they also included studies on the rusting of iron. Young scientists usually need role models. I regard it as a blessing that I found in Heinrich Wieland, my senior by 43 years, a friend who took an almost paternal interest in me.

Not a Rose Garden

Wieland used to give the introductory courses, which included experimental demonstrations, and as a graduate student I was responsible for the experiments. Incidentally, I presented my first lecture in organic chemistry to more than 300 students at the age of 23. This was in 1943, after an air raid had stopped suburban traffic so that Wieland could not reach the institute. On the other hand, he insisted that not a single lecture of the compact course should be canceled.

Research in Munich was more and more impeded by air raids, even before the laboratory built under Justus von Liebig, Adolf von Baeyer, and Richard Willstätter was destroyed in 1944. After the war, the chemistry department was revived in various colonies scattered over Munich's rural environment. In a wooden shack in Weilheim (50 km south of Munich), provisionally equipped for laboratory work, I instructed 12–15 undergraduates in organic chemistry, beginning in 1946; soon some graduate students joined the group. In those years, laboratory courses and research were daily exercises in improvisation, a victory of mind over matter.

The German chemical industry—BASF, Bayer, and Hoechst in particular—supplied chemicals. In response to Otto Bayer's initiative, the *Fonds der Chemischen Industrie* (Chemical Industry Fund) was established to support education and research at the universities in 1950. As one of the beneficiaries of that fund to this very day, I am sincerely grateful to those who established it and to those who kept up the tradition. I feel no less obliged to the *Deutsche Forschungsgemeinschaft* (German Research Association), which financed much of my research.

In the postwar years, the condition of the departmental library was desolate. The collection of monographs was lost; the journals had been saved, but access to them was difficult. Moreover, the gaps in acquisitions due to the war were painful. Our first contact with the new issues of the *Journal of the American Chemical Society* and *Chemical Abstracts* was made at the America House in Munich, where new journals were put on display. In the late 1940s we received some back issues of these journals from R. B. Woodward. A few years later, further volumes followed as gift of the U.S. State Department on the

Once this was the laboratory in which Justus von Liebig, Adolf von Baeyer, Richard Willstätter, and Heinrich Wieland performed classic research work. The ruins of the chemistry building, University of Munich, are shown after World War II (1945).

recommendation of Richard Arnold, who served as science adviser to the high commissioner of Germany.

Predicament is an effective teacher of the art of economizing. In stark contrast to the postwar era, hundreds of millions of German Marks were spent in the 1970s to build gigantic institutes for training and research. Perhaps these institutes reflect the *Zeitgeist* (genius of the period) in the same way as the great cathedrals marked the Gothic and Renaissance periods. It is doubtful, however, whether future generations will be as keen on preserving these "temples of science" as we are now on guarding the heritage of the past.

Aiming at New Goals

During the postwar years, Linus Pauling's *The Nature of the Chemical Bond* (1939) and George W. Wheland's *Theory of Resonance* (1944) were revelations to me. Moreover, I was greatly impressed by Hans

Meerwein's studies on onium ions and cationic rearrangements, as well as by C. K. Ingold's excursions into the realm of substitution mechanisms. Unifying concepts capable of bringing order into the chaos of facts were expected to spring from the exploration of reaction mechanisms. I felt attracted by this field and anticipated more important insights from this approach than from the structural elucidation of further alkaloids.

Of course, I was not aware of the fact that many young scientists, worldwide, were beginning to direct their research activities toward mechanistic and physical organic chemistry. These areas were to play a dominant role for the two subsequent decades. Such fashions in science appear to reflect compelling inherent needs in the development of a subject. They act as motors by triggering highly concentrated efforts. A current example is the rush of scientists to study fullerenes. New modifications of elementary carbon must be considered fundamental developments. The rapid spread of research on fullerenes was recently tracked by application of an *epidemiological* model.[4c]

H. Wieland kindly consented to independent ventures on my part. First these ran parallel to my work on alkaloids, but soon they superseded it. My habilitation thesis (1947) dealt with angular versus linear fusion of heterorings starting from β-substituted naphthalenes and quinolines.[5,6] Some exceptions to the superiority of angular annellation proved mechanistically more instructive than a shipload of further confirmations.

Climbing the Academic Ladder

In 1949 I accepted an offer from the University of Tübingen and served as professor extraordinarius in Georg Wittig's institute until 1952. Tübingen was largely undestroyed. The chemistry building, with its fortress walls dating back to the 1890s, could hardly accommodate the great number of students who had returned from war and captivity. I admired Wittig's artistic research style and his unusual gift of observation. Moreover, I closely cooperated with my small research group in the field of nitrosoacylamides (page 15) and, relying on my own hands, studied the coupling of aromatic with aliphatic diazo compounds (page 28). Much midnight oil went into the preparation of my courses on advanced organic chemistry. I experienced the truth of the old maxim that the main beneficiary of lectures is the lecturer.

Junior faculty members were in high demand because of wartime losses. In 1951 I was offered a professorship at the University of Marburg as the successor of Hans Meerwein. Soon I was also offered a position at Erlangen to succeed Rudolf Pummerer, even though, at the age of 31, I lacked the high number of publications that is expected of candidates today. In Germany, universities fall under the responsibility of the states. While I was still negotiating with the ministers of education, a third offer arrived from Munich. The traditional chemistry building still being a shambles, Richard Kuhn and Clemens Schöpf had declined to take over Wieland's chair. Because the new Bavarian minister of education gave reconstruction high priority, the challenge appeared reasonable, and I accepted. I well understood that I received the chair at Munich, rich in tradition, not as a reward for scientific merits but rather as a trust for the future.

In the spring of 1952 I thus moved to Munich with my group and resumed work in temporary laboratories. H. Wieland had made a deal with Karl von Frisch, the famous zoologist: The chemistry department was to help in procuring bricks and cement to repair the zoology building and add an additional floor to it; in return, the chemists were

A great moment for the new Institute of Organic Chemistry, University of Munich: The first wing of the chemistry building neared completion in 1955.

allowed to occupy two floors of the zoological institute until their own building was finished. Much effort went into planning. The minister of education kept his word, and early in 1956 the organic chemistry group took possession of brand-new laboratories. Four years later the lecture halls, named after Justus von Liebig, Adolf von Baeyer, and Richard Willstätter, were completed.

The building at Karl Street, planned in 1952, contained 80 m^2 of floor space for physical organic chemistry laboratories. That was plenty of room for rate and dielectric measurements, a polarimeter, a UV spectrophotometer, and (installed a few years later) an IR instrument. However, it was woefully inadequate for all the equipment made indispensable by the rapid development of the following decades: gas chromatography, mass spectrometer, nuclear magnetic resonance spectrometer, high-pressure liquid chromatograph, and computer terminal. The perpetual need for remodeling created widespread headaches, not only in our institute.

Lecturing

The introductory courses of Paul Pfeiffer and Heinrich Wieland, to which I had been exposed, were very powerful in kindling enthusiasm. As a professor, I consequently tried to maintain the tradition. I presented the introduction to organic chemistry every summer and found great pleasure in adding to the inventory of strategically selected demonstration experiments. I pondered over didactics, level, and choice of material, as these lectures were also attended by students of medicine, pharmacy, and biology. The variety of student backgrounds required that the instruction must be broad. The beginner must be made aware of the extent to which organic chemistry pervades our lives.

The Munich tradition of lecture demonstrations goes back to Justus von Liebig and Adolf von Baeyer. Liebig's public evening lectures were often attended by the royal family. Some of Baeyer's demonstrations are still in the repertory (e.g., the temperature limits of nitroglycerol detonation and the bromine-catalyzed explosion of silver acetylenide). Students are fond of acoustical effects; however, some prefer optical effects because loud bangs would interfere with their morning doze.

Throughout my academic life I have been seriously involved in teaching. This commitment limited travel during the summer semesters, when I dealt with *Organische Experimentalchemie* (experimental organic chemistry). The advanced courses that I presented during the winter

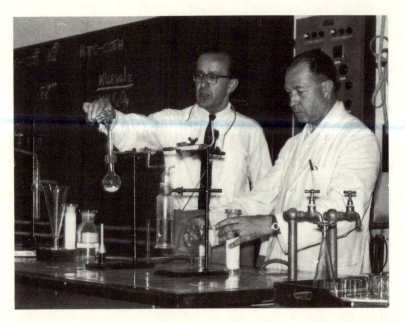

During lecture demonstrations in 1955 I was assisted by technician J. Unterreit-meier (right) in the provisional lecture hall. The permanent auditorium (Justus-von-Liebig-Hörsaal) seats 700, and the reagent tube in the demonstrations has been replaced by a 20-L beaker.

semesters offered more flexibility. Consequently, my lecture trips and guest professorships were scheduled during the winter terms and the breaks between terms. Today the emphasis I placed on teaching seems a bit old-fashioned; it may be a legacy of Heinrich Wieland.

Early Scientific Contacts and Research

Foreign scientists established personal bonds with German colleagues rather soon after the war. It was a delight to see our isolation overcome by guests who visited Munich and lectured in the early 1950s: Roger Adams, Richard Arnold, Paul D. Bartlett, Donald Cram, William von E. Doering, Arne Fredga, William S. Johnson, Linus Pauling, John D. Roberts, and Arthur Stoll. James B. Conant, president of Harvard University and at that time the United States high commissioner in Germany, visited the Munich Institute in 1953 and discussed chemistry. In 1957 Paul Bartlett spent a sabbatical in Munich; Joseph Bunnett, Philip Skell, Harold Zeiss, and Aksel Bothner-By followed a few years later.

With the discoverer of tropylium ion and carbenes, William von E. Doering (left), in Munich in 1958. (Photograph by Hermann Stetter.)

Roger Adams (1899–1971) at the University of Illinois, Urbana, in 1955. He was a postdoctoral assistant in Berlin from 1912 to 1913. After World War II Adams visited West Germany searching for promising young chemists.

Research is strongly promoted by personal contacts. I was to experience this truism when I toured the United States as a Rockefeller fellow for 3 months in 1955, visiting many chemistry departments and establishing numerous connections, some of which developed into friendships.

The dream of a young scientist on his first trip to the United States is an encounter with the masters, here the Harvard brain trust: Robert B. Woodward, George B. Kistiakowsky, and Paul D. Bartlett in 1955.

Research projects on intramolecular Friedel–Crafts acylation, on the occurrence of arynes in nucleophilic aromatic substitution, and on the chemistry of azodicarboxylic ester (all discussed later), which had been started in Tübingen, came to fruition in Munich. In parallel efforts, we studied the thermolysis of azides, the pentazole problem, and the *cis–trans* isomerism of lactams and lactones.

The experimental sciences require the cooperation, skill, and diligence of many co-workers to bring ideas to life. With gratitude I recall the enthusiastically active postwar generation of students in the 1950s and 1960s. It appears that a common struggle with hardships unites, whereas most of the benefits that come with prosperity detract from the essentials. Soon the research group grew until it included around 20 members.

Academic Freedom: Fact and Fiction

Academic freedom is, of course, a precious good in theory. Its practice, however, is somewhat disillusioning, probably not only at German universities. If we compare a scientist's time to the contents of a wine bottle, most of the contents would be administrative chores, committee meetings, and exams; the time available for research corresponds to the final sip—and it is still rewarding enough.

Journalists are fond of envisioning scholars in an "ivory tower", far from the maddening crowd. By contrast, I remember a day in July 1973 when a horde of bearded fellows disrupted my lecture to spread the message of social revolution to about 400 freshmen. Even though the silent majority increasingly sympathized with my efforts to save the lecture, it took a 35-minute shouting match before the intruders left. This final battle ended 4 years of student rebellion in our department, 4 years of stemming the tide, frantic faculty discussions, even colleagues' opportunism. What ivory tower?

Naturally, research projects tend to branch out in various directions, but not all ramifications can be reported here without obstructing the clear view. The thicket that soon results from the germination of a series of seminal ideas requires lopping and thinning to yield a viable forest. The criterion in this process of trimming is significance.

But what is significant? Here personal judgment enters the scene, and likewise that of peers, reviewers, and audiences. The scientist interacts with a community, and the decision about what is significant is largely based on consensus. We are again talking about fashions, their beneficial influence, and their dangers. Both jumping on the bandwagon and catapulting overlooked fields into the limelight are reactions of the individual scientist, determined largely by temperament and degree of self-confidence.

Old foundations one should venerate, without giving up the right to start the process of laying them all over again.

Johann Wolfgang von Goethe

The reader is invited to stroll through the "cultivated forest"—admittedly with some self-consciousness on the part of the author. Major research lines will be presented without any claim of completeness. For more personal recollections see pages 187–245.

Old and New Reactions of Diazo Compounds

The aromatic diazonium ion has fascinated generations of chemists: molecular nitrogen forced into the onium state, metastability despite a high energy level, a wealth of reactions equally significant for laboratory and industry.

Radical Phenylation and the N-Nitrosoacetanilide Puzzle

H. Wieland's interest in reaction mechanisms—a side field—led to a series of publications, *Über das Auftreten freier Radikale bei chemischen Reaktionen*[7,8] (On the Occurrence of Free Radicals in Chemical Reactions), in which he dealt with the decomposition of diacyl peroxides and arylazo-triphenylmethanes. It was shown in 1934 that phenylazotriphenyl-methane in refluxing CCl_4 provided chlorobenzene and 1,1,1-trichloro-2,2,2-triphenylethane instead of tetraphenylmethane. I was convinced that the application of spectroscopic and kinetic techniques would lead to deeper understanding of this and many other observations.

In 1937 Hey and Waters[9] attributed phenylations of aromatic solvents by benzenediazohydroxide or N-nitrosoacetanilide (1) to the free phenyl radical. The evolution of N_2 from 1 obeyed first-order kinetics, and the rate constant was nearly independent of the nature of the solvent. Grieve and Hey[10] considered the dissociation of benzenediazo-acetate (2) into radicals and N_2 as rate-determining. In accordance with Bamberger,[11] the covalent 2 was presumed to occur in rapid equilibrium

15

with **1**. This kinetic feature was assessed by Hey and Waters[12] as "crucial experimental evidence for the existence of free aryl radicals".

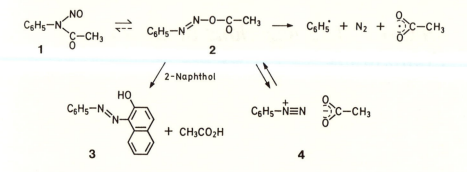

Inconsistencies prompted my investigation in 1947. A photometer with visual comparison of light intensities was available at the Weilheim Laboratory, and I constructed a thermostat. In the presence of 2-naphthol, the generation of radicals from **1** was suppressed in favor of the quantitatively formed azo dye **3**. The rate of azo coupling measured by photometry or gravimetry was first order in **1** and zeroth order in 2-naphthol. It was not the homolysis, but rather the *intramolecular acyl migration*, **1** → **2**, that controlled the rate. In the absence of 2-naphthol, **1** disappeared with the same half-life, as volumetry of N_2 evolution showed.[13] My first co-worker in that area was Gabriele Horeld.

The reaction of **1** in acetic acid afforded phenyl acetate as a product of ionic decomposition. Electrical conductivity of the solution and capacity of azo coupling went parallel, indicating that electrolyte **4** was responsible.[14]

There was more to learn from the radical reactions of **1**. Because the acyl shift **1** → **2** is slow, the phenyl radical must originate from a fast subsequent step. The acetoxy radical expected as the second fragment of **2** was known to break down into $CH_3^{\bullet} + CO_2$. However, CO_2 was not formed in the decomposition of **1** in aromatic solvents; instead, the production of 95% acetic acid suggested the overall reaction:

$$C_6H_5-N=N-O-CO-CH_3 + ArH \rightarrow$$
$$C_6H_5-Ar + CH_3-CO_2H + N_2 \qquad (1)$$

How were we to overcome the dilemma? We postulated the following scheme with a single electron transfer (SET) from benzene to benzenediazoacetate (**2**), thus inducing a fragmentation into phenyl radical, acetate anion, and N_2.[13] SET was not so much in fashion in 1949 as it is today.

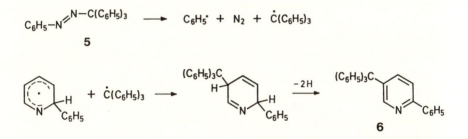

$$C_6H_6 + C_6H_5-N\diagdown^{N-O-C-CH_3}_{O} \longrightarrow \langle\rangle^{+\cdot} + C_6H_5^{\cdot} + CH_3-CO_2^- + N_2$$

$$\langle\rangle-C_6H_5 + CH_3-CO_2H + N_2 \longleftarrow$$

The homolysis of phenylazotriphenylmethane (5) was free of ambiguity. Thermolysis in benzene at 50 °C and photolysis at 20 °C yielded 58% biphenyl and 57% triphenylmethane; triphenylmethyl radical was trapped by iodine to the extent of 77%.[15] Besides phenylpyridines and triphenylmethane, the interaction of 5 with pyridine yielded 6, which emerged from a radical recombination.[15,16] These observations were made by Herbert Nakaten, fellow student and friend, who joined me at Weilheim after returning from war captivity.

$$C_6H_5-N\diagdown^{N-C(C_6H_5)_3} \longrightarrow C_6H_5^{\cdot} + N_2 + \overset{\cdot}{C}(C_6H_5)_3$$

5

As a result of SET, the phenyl radical attacks a benzene radical cation in the preceding scheme in contrast to $C_6H_5^{\cdot}$ + benzene in the reaction of 5. When we received our first IR spectrophotometer, the isomer ratios of chlorobiphenyls obtained from chlorobenzene and various sources of the phenyl radical (1, 5, and dibenzoyl peroxide) were compared and found identical: 57% *o*-, 27% *m*-, and 16% *p*-isomer. Analogously, the three phenyl generators converted naphthalene into 1- and 2-phenyl derivative in an 80:20 ratio.[17] Hey et al.[18] observed isomer ratios of phenylpyridines that were likewise independent of the phenyl source. These results supported identical phenylation mechanisms for 1 and 5,[19] and the SET dream vanished.

New observations in the 1960s—without our participation—brought the N-nitrosoacetanilide puzzle close to a solution. Suschitzky et al.[20] found 4-acetoxybiphenyl along with 4-fluorobiphenyl, when N-nitroso-4-fluoroacetanilide was decomposed in benzene. This result pointed to the intermediacy of 4-fluorobenzenediazonium ion. My former student, C. Rüchardt and co-workers[21] isolated benzenediazonium-4-chlorobenzoate when N-nitroso-4-chlorobenzanilide was allowed to isomerize in CCl$_4$; the covalent diazoesters of type 2 *ionize spontane-*

ously in nonpolar solvents. The conclusion is accepted today: The benzenediazonium ion is the precursor of the phenyl radical. Radicals **7** and **8** were observed by electron spin resonance spectroscopy in decomposing solutions of **1**. Various chain reactions were proposed by Rüchardt,[22] Perkins,[23] and Cadogan.[24] A major contributor may be the chain in which the adduct radical **9** donates the electron to the benzenediazonium ion and the acetate deprotonates.

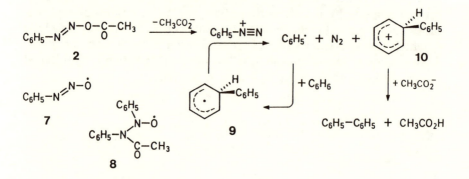

1,3-Acyl Migration and Deacylation
of *N*-Nitrosoacylamides

The solution of one problem usually generates a bevy of new ones. The inexperienced young scientist often lacks the willpower to resist the temptation of dealing with a new problem while working on the first one.

What is the mechanism of the isomerization **1** → **2**? *N*-Nitroso-*N*-acyl derivatives of primary aromatic and aliphatic amines, **11**, isomerize by 1,3-acyl shift; the covalent *trans*-diazoesters **13** enter into fast secondary reactions. In a broad variation of R and R' in **11**, rates of acyl migration were measured by azo coupling of **13**, R = aryl,[25] and by volumetry of N_2 extruded from **13**, R = alkyl.[26]

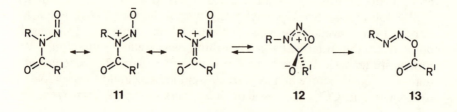

The isomerization rate of N-nitrosoacetanilide (**11**, R = C_6H_5, R' = CH_3) is only slightly influenced by p-substituents, electron-releasing or -attracting.[25] In **11**, R' = C_6H_5, a 1600-fold rate increase in the sequence R = methyl < ethyl < isopropyl < cyclohexyl indicates steric acceleration. Acyl variation allowed the same conclusion. The rate constant k_1 increased 7400-fold for **11**, R = $C_6H_5CH_2$, in the range between R' = H and *tert*-butyl; the k_1 values for propionyl and chloroacetyl were similar.[26] Here I relied on the careful work of Hans Reimlinger.

The resonance structures of **11**, R = aryl, teach that nitroso, acyl, and aryl groups compete in decreasing order for the electron pair of the central nitrogen. The planar *trans,trans* configuration of type **14** is preferred to *cis,trans* because of compensation of partial dipole moments stemming from the nitrosamine and carboxamide resonance. In the nitrosolactam **15** another conformation is "frozen" by ring closure. My co-worker Josef Reinertshofer (1923–1959) and I were struck by the profound influence of the configuration on dipole moment, volatility, and solubility (Table I).[27]

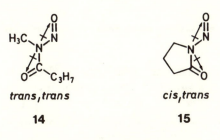

trans,trans   cis,trans

14   **15**

Table I. Physical Properties
of Nitroso–*N*-methylbutyramide (**14**)
and Nitrosopyrrolidone (**15**)

Property	14	15
Dipole moment (Debye, C_6H_6)	0.92	4.58
Boiling point (°C, 14 torr)	60	123
Solubility (g/L, 25 °C)		
in cyclohexane	∞	0.82
in water	3.8	129

Although the rate sequence does not resemble that of alkaline ester hydrolysis, a carbonyl addition to give **12** was assumed to determine the rate of the $N \rightarrow O$ acyl shift. This carbonyl addition requires the *cis* configuration of the nitrosamine part and twisting of the carboxamide's CN bond. Bulky R and R' in 11 disturb the planarity and decrease the activation energy of acyl migration, because 12 suffers less from varying steric demands of R and R'. The isomerization rate of *N*-nitrosolactams[27] and *N*-nitrosobenzlactams[28] show a dependence on ring size that fully agrees with 11 → 12 as the rate-determining step.

N-Nitrosoacylamides acylate amines but not alcohols, thus resembling acyl azides in acylating activity. The acyl transfer from *N*-nitrosoacylanilides (11, R = C_6H_5) to piperidine is magnitudes faster than the intramolecular acyl shift, 11 → 13, and the reactivity sequence (formyl:acetyl:chloroacetyl = 116:1:160)[29] parallels that of alkaline ester hydrolysis (213:1:760).

The preparation of diazoalkanes calls for generation of alkanediazonium ions in basic medium. Hans von Pechmann[30] discovered diazomethane upon treating *N*-nitrosomethylurethane with methanolic KOH. Nucleophilic deacylation of *N*-nitrosoalkylacylamides (in dozens of variations) has been the most popular pathway to diazoalkanes.[31] The carcinogenicity of reactant and product was not known in the early 1950s.

Kinetic arguments are crystal clear and cannot be overturned. I developed a predilection for chemical kinetics, although I had never heard lectures on that topic. In the postwar years in Weilheim (until 1949), no literature on kinetics was available. It is unproductive to derive long-known integrated rate equations for complex systems, but it was a lesson for a lifetime. I often notice in discussions of rate phenomena that many fellow chemists are not aware of the laws of formal kinetics (e.g., the apparent acceleration of both branches in parallel reactions).

My passion for chemical kinetics stems from these early projects; no other method could have provided so much insight. I became tired of explaining kinetic techniques and proper evaluation of data to every new student. In 1954 I wrote an article, *Ausführung kinetischer Versuche* (Performing Kinetic Experiments), for Houben-Weyl's *Methoden der Organischen Chemie* (Methods of Organic Chemistry).[32] My co-workers still like to use this concise guideline because it emphasizes experimental aspects.

Experience in chemical kinetics has taught me to prefer big rate effects to small ones. This choice is not meant as a joke, like choosing between a lot and a little money. Our understanding of rate phenomena is fragmentary, and our theoretical tools are blunt. Big rate ranges offer a better chance of reaching reasonable conclusions.

In the chemistry of high-energy intermediates, methods of kinetic competition are indispensable. Many applications will be described here. One of my favorite examples involves the competition of uni- and bimolecular reactions. In such a competition the higher temperature coefficient is anticipated for the unimolecular reaction. The making of one molecule out of two is always burdened by a negative entropy of activation. The different temperature dependencies must be weighed by the experimenter in determining the reaction temperature. For example, cycloadditions furnishing labile adducts should be run at as low a temperature as feasible (page 90). Moreover, the simple consideration prompted the clarification of the Schönberg reaction (page 110).

Fifty years ago many distinguished chemists were not yet accustomed to separating kinetic and thermodynamic phenomena or to visualizing a chemical reaction in terms of energy. Two-dimensional energy profiles are highly artificial as a conceptual tool, and most students are unaware of their shortcomings. Nevertheless, I am slightly frustrated when discussion partners today are not familiar with thinking in energy profiles.

Aromatic Diazoesters: Azo Coupling

In 1870 August Kekulé discovered azo coupling. The covalent benzenediazohydroxide was long regarded as the active species until experiments by P. D. Bartlett (1941), R. Pütter (1951), and H. Zollinger (1952–1953) attributed this role to diazonium ion + phenoxide ion.[33]

In 1949 we did not dare to assume that benzenediazoacetate (2) might ionize in nonpolar solvents to give 4.[13,14] Instead, we described the electrophilic azo coupling in nonpolar medium by a concerted four-center process of 2 with 2-naphthol.[34] In 1954 pK measurements by Zollinger et al.[35] revealed that benzenediazohydroxide is not a stable species in aqueous medium, which was all the more reason to doubt the stability of the covalent diazoacetate 2. Today the four-step sequence for the azo coupling of 2 in nonpolar media appeals to me as more likely: ionization 2 → 4, deprotonation of naphthol by the acetate anion, ion recombination, and prototropy.

In 1908 Jacobson and Huber[36] reported the formation of indazole (17) from N-nitrosobenz-o-toluidide (16, R = C_6H_5) in warm benzene. Under optimal conditions we obtained 95% of 17 and submitted the conversion of o-toluidine via 16, R = CH_3, into indazole to *Organic*

Syntheses.[37] Interaction of toluene-2-diazonium chloride with tetrameth-ylammonium acetate in CHCl$_3$ afforded 87% of **17**. In the presence of 2-naphthol, intermolecular azo coupling won over the intramolecular process as observed by Herbert Nakaten (Ph.D. 1951).[38]

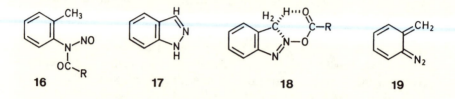

16   **17**   **18**   **19**

    We were prejudiced and proposed **18** as the transition state of a concerted indazole formation in nonpolar medium (i.e., **19** was assumed to be a hyperconjugated resonance contributor). In 1978 Rüchardt et al.[39] elegantly demonstrated the occurrence of a type-**19** intermediate by the formation of a racemic *3H*-indazole from an optically active deriva-tive of **16** (−CH(CH$_3$)CO$_2$CH$_3$ instead of CH$_3$).

    The carboxylate anion is capable of deprotonating the benzenedi-azonium ion in the nonpolar medium. This pathway to *benzyne* was an exciting new chapter of *N*-nitrosoacetanilide chemistry connected with the names of Cadogan[40,41] and Rüchardt.[42,43] Ironically, research on nitrosoacylamines and benzyne chemistry (page 65) ran side by side in the Munich laboratory in the 1950s without anyone recognizing the connection.

Alkanediazoesters: Source of Diazonium Ions and Carbocations

No radicals occurred in the decomposition of *N*-nitroso-*N*-*alkyl*acyl-amides via *trans*-diazoesters. *N*-Nitrosobenzylbenzamide (**20**) in ben-zene at 70 °C afforded 86% of benzyl benzoate (**23**); the reaction in toluene allowed the isolation of isomeric benzyltoluenes. In refluxing methanol, **20** gave rise to benzyl methyl ether (**22**) and ester **23** in 66:34 ratio.[44] The formation of 31% of *N*-acetylbenzamide (**26**) and 55% of ester **23** in acetonitrile likewise indicated benzylation of the solvent, here via **25**.[45]

    Although it seemed more daring than in the aromatic series, we did postulate the alkanediazonium ion pair **24** in 1953.[46] We saw the connection between our problem and the worldwide debate about nucleophilic substitution, which may still be rated as the deepest inroad

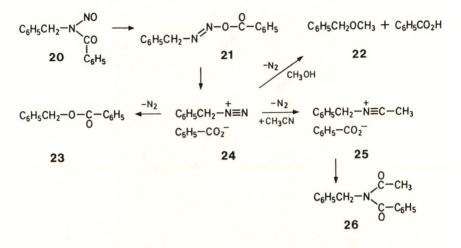

in reaction mechanisms so far. Solvolysis developed to an American passion under the leadership of Saul Winstein, and ion pair became a catchword. I recall enlightening discussions with Winstein in Munich, as well as at University of California at Los Angeles; they were not even interrupted by our morning swim in his pool in Bel Air (Los Angeles). The Cram–Winstein evening seminar was notorious as a "blood bath"; allegedly, some speakers did not get far beyond the prelude in their presentations. Was it through the hosts's politeness that I survived the event unscathed? After a couple of interruptions during the presentation I shot back unmoved and was allowed to finish my talk before a constructive discussion started.

N-Nitrosoacylamides offered the unique chance of studying solvolysis phenomena at a low stationary concentration over the full scale of organic solvents. The dialog was opened, with other research groups [e.g., E. H. White (1955) and A. Streitwieser (1957)] who used the same access to alkanediazonium ions.

Rearrangements are indicative of carbocationic mechanisms. N-Nitrosopropylbenzamide in benzene at 70 °C produced an ester that contained 2% of isopropyl benzoate. In 60% aqueous dimethylformamide (DMF) at 70 °C, benzoic esters and alcohols were formed; the propyl:isopropyl ratio amounted to 91:9 in the ester and 67:33 in the alcohol, respectively. Similar mixtures of propyl and isopropyl alcohol (60% DMF, 0 °C) resulted from 1-diazopropane and benzoic acid or perchloric acid, as well as from propylamine + HNO_2, as elucidated by C. Rüchardt.[47] The same intermediates must be involved. In the acid-catalyzed rearrangement of 2-anisyl-2-phenylethanol, anisyl migrates 21 times faster than phenyl; in the reaction of the amine with nitrous acid, the migration ratio dropped to 1:4.[48] We observed 1:3 as migration

Christoph Rüchardt was one of my first graduate students in Munich (Ph.D. 1956) after my move from Tübingen. He became a professor in Münster in 1968 and in Freiburg in 1972. His interest in radical chemistry was awakened when he worked for Paul D. Bartlett as a postdoctoral fellow. His studies on bond dissociation energies and substituent effects are widely appreciated. (This photograph was taken in 1954.)

ratio in the nitrosoacylamide decomposition in nonpolar solvent.[47] Why is the selectivity smaller in the reactions via diazonium ion?

We argued in 1956 that N_2 extrusion from alkanediazonium ions constitutes the only exothermic ionization process leading to carbocations, at least to the stabilized ones. The exothermicity would render the transition state more reactant-like; neighboring group and solvent participation may play a role, but not a mandatory one.[47] However, the dissociation energy of methanediazonium ion in the gas phase, +38 kcal mol^{-1}, was measured in 1972 by ion cyclotron resonance spectroscopy.[49] Ab initio calculations of the free energies for $R\text{-}N_2^+ \rightarrow R^+ + N_2$ provided +28, +1, and 0 kcal mol^{-1} for R = methyl, ethyl, and propyl, respectively.[50] Resonance-stabilized carbocations might well be formed exothermally.

The nebulous concept of "hot carbonium ions" has been applied since 1958 in order to explain differences between products of alkyldiazonium decomposition and those of alkyl halide solvolysis.[51] Carbocations stemming from diazonium ions were supposed to have excess energy and to react before thermal equilibrium with the environment is achieved. The smaller role of inter- and intramolecular nucleophilic assistance in the N_2 extrusion, as pointed out, is sufficient clue. The degenerate rearrangement in a rigid tricyclic cation served as a test and showed no dependence on the precursor.[52]

To which extent are *diazoalkanes* involved in the reactions of *N*-nitrosoacylamides via alkanediazonium carboxylates? Both *N*-nitrosomethylbenzamide (**27**) + 4-nitrobenzoic acid and the pair **31** + benzoic acid produced mixtures of methyl benzoate and methyl 4-nitrobenzoate at 90 °C. Numerical evaluation by Hans Reimlinger revealed free competition, 4-nitrobenzoic acid being 2.6 times more successful than benzoic acid in overall ester formation. The interaction of diazomethane with a mixture of the two acids (−10 °C) provided virtually the same competition constant. Either both ion pairs **28** and **32** pass the diazomethane stage **29** or an ion pair **30** with the monoanion of a dimeric carboxylic acid is responsible for the competition.[44]

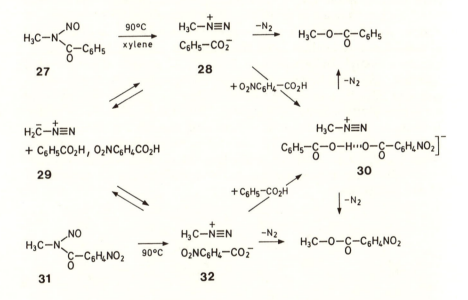

The results of H–D exchange established the involvement of diazomethane. The reaction of *N*-nitrosomethylbenzamide in toluene at 70 °C in the presence of an excess of C_6H_5–CO_2D afforded methyl benzoate with more than 1 atom equiv of D. Diazomethane must be passed several times, without full equilibration of H–D exchange. Analogously, diazomethane and 2 equiv of C_6H_5–CO_2D furnished methyl benzoate with more than 1 atom equiv of D.[45]

Lower D incorporations in alkyl benzoates were observed for other *N*-nitrosoalkylbenzamides. Supplementary data came from reactions in the presence of $C_6H_5-^{14}CO_2H$. The competing formation of alkenes and benzoic acid from the alkanediazonium ion pair and the slow release of acid bound in the nitrosoalkylbenzamide led to a rather complex equation for quantitative evaluation. The maximal passage through diazoalkane amounted to 97% for methyl, 72% for benzyl, 42% for isopropyl, and 39% for α-phenylethyl.[45]

Heinz Stangl was not only the ski ace of the group, but was also successful in establishing the deuterium analysis by the somewhat touchy procedure of Schönheimer and Rittenberg: density measurement of the combustion water by the falling drop method. The electrometric analysis of radiocarbon in CO_2 after wet combustion also required great skill. The simpler scintillation methods had not yet been introduced in the 1950s.

Four reactions compete for the alkanediazonium ion pair: reversible deprotonation giving diazoalkane + benzoic acid; nucleophilic attack by benzoate anion producing alkyl benzoate or alkene + benzoic acid; or spontaneous N_2 loss generating the carbocation. α-Phenylethyl is a stabilized carbenium ion. The formation of ester **36** from *(S)-N*-nitroso-α-phenylethylbenzamide **(33)** at 35 °C proceeded with overall *retention*: 43% enantiomeric excess (ee) in benzene, 34% in nitromethane, and 27% in acetonitrile. Racemization accompanies the pathway via α-phenyldiazoethane (39% above). Thus, the remaining 61% of diazonium ion pair **34** in benzene yielded ester **36** with a retention of 71% ee. This is a fair value for the oriented carbenium ion pair **35** in benzene. An S_N2-type ester formation from **34** should give rise to inversion.[53] In these stereochemical studies I enjoyed the cooperation of Christoph Rüchardt.

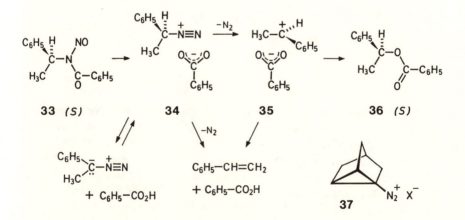

The reaction of **33** in 50% aqueous tetrahydrofuran (THF) furnished 27% ester **36** with 54% ee and 37% of α-phenylethanol with the lower retention of 8% ee. In the presence of acetic acid in benzene, **33** likewise provided benzoate **36** with higher retention than α-phenylethyl acetate (i.e., the orientation vanishes with disturbance of ion pair **35**).[53]

Nonstabilized primary alkanediazonium ions still react by nucleophilic attack. In 1957, Streitwieser et al.[54] described the reaction of optically active [1-D$_1$]butylamine with nitrous acid in acetic acid; 69% ee of inversion in the butyl acetate suggested direct displacement at the *primary* alkyldiazonium ion.

The preparation of the methanediazonium ion from diazomethane and fluorosulfonic acid in SO$_2$ClF by McGarrity and Cox[55] is a landmark of recent progress. The cationic species was stable in the superacid medium at −120 °C and yielded methyl fluorosulfonate + N$_2$ at −80 °C. Allyldiazonium ions served as models for Kirmse's careful differentiation[56] between unassisted and nucleophile-assisted N$_2$ loss. The nitrogen extrusion from cyclopropanediazonium ions requires assistance by electrocyclic ring opening. Nortricyclenediazonium ion (**37**) is barred from disrotatory ring cleavage and from nucleophilic rearside attack; the poor quality of the nortricyclyl cation gives sufficient lifetime to the diazonium ion **37** for azo coupling with naphthoxide.[57] The distinction of Wolfgang Kirmse's mechanistic work may be emphasized.

We did our research under conditions very different from those of today. From 1947 to 1959, the *N*-nitrosoacylamides project profited

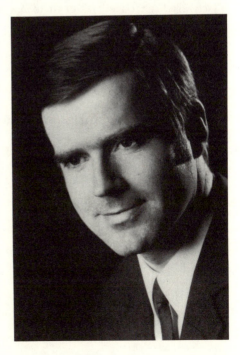

Bernd Giese's fine mechanistic study on the addition of amines to acetylenecarboxylic esters (Ph.D. Munich, 1969) is not covered in this review. After some years at BASF AG he habilitated at the University of Freiburg in 1977. Splendid research on the reactivity–selectivity principle, on radical additions to alkenes, and on radical reactions in organic synthesis attracted international attention. In 1978 Giese became professor at Darmstadt and moved to Basel in 1989.

from the increased support and improved equipment associated with each move, from Weilheim to Tübingen and further to Munich. The work described fell into the pre-NMR age, often mocked now as a kind of prehistory of chemistry. Many have nearly forgotten, if they ever knew, that IR band intensities allow quantitative analysis with higher precision than NMR spectroscopy. Admittedly, IR analysis requires more experience, skill, and time.

With seven graduate students involved over a period of 12 years, N-nitrosoacylamides were not a major research effort. I was (and still am) fond of parallel activities in several, sometimes unrelated, research fields. A project may go through a stagnation phase in its evolution; when this happens, encouragement is drawn from other research programs bearing fruit.

Coupling of Aromatic Diazonium Ions with Aliphatic Diazo Compounds

The diazonium ion, $-N_2^+$, is the strongest electron-attracting group. The electrophilic aromatic diazonium ion owes its existence to a resonance stabilization that the alkanediazonium ion achieves by deprotonation. Diazoalkanes are part of Munich's tradition: Theodor Curtius[58a] discovered diazoacetic ester in 1883 and Hans von Pechmann[30] prepared diazomethane in 1894. In 1955 I attempted to order and classify the colorful reactivity of diazoalkanes in a review.[58b] A novelty was included: the coupling of the electrophilic aromatic diazonium ion with the nucleophilic diazoalkanes.

4-Nitrobenzenediazonium chloride (38) combined with ethyl diazoacetate or diazoacetophenone to give the α-chlorohydrazones 41, R = $CO_2C_2H_5$ and COC_6H_5. When diazomethane was passed into the methanolic solution of 38, saturated with lithium chloride, the hydrazone of formyl chloride, 41, R = H (80%) was isolated by Hans-Joachim Koch (Ph.D. 1954).[59] Product 39 of initial coupling loses N_2 and the cationic intermediate 40 captures the chloride anion, the tautomerization 42 → 41 being the concluding step.

Slow introduction of a methanolic solution of 38 into ethereal diazomethane (i.e., the reverse addition) furnished 4-nitrophenylcyanamide (43, 24%).[59] A $^{15}N^\beta$ label in the diazonium salt 38 appeared in the nitrile group of 43,[60] and the antiaromatic 2H-diazirine derivative 44 is a conceivable intermediate.

The tautomerization step, 42 → 41, is blocked in the analogous coupling products from 38 and diphenyldiazomethane or 9-diazofluo-

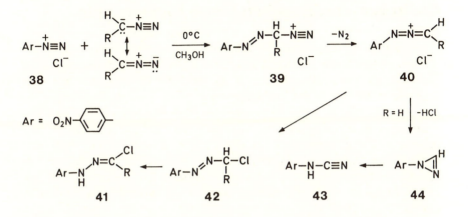

rene; methanolysis of the intermediate chloride **42** gave rise to α-aryl-azoethers like **45**, Ar = $C_6H_4NO_2$.[59]

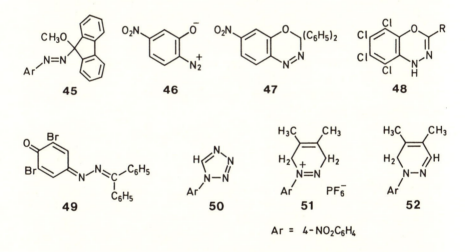

Ar = 4-$NO_2C_6H_4$

Ion recombination **40** → **42** was replaced by ring closure when 2-diazophenoxides bearing electron-attracting substituents like **46** were used by Rudolf Fleischmann (Ph.D. 1959). Diphenyldiazomethane or diazofluorene were converted to yellow 2*H*-benz[*e*]-1,3,4-oxdiazines [e.g., **47** (94%)]. Reactions with monosubstituted diazomethanes furnished 4*H*-benzoxdiazines (e.g., **48**, R = C_6H_5, from tetrachloro-2-diazophenoxide and phenyldiazomethane, and **48**, R = $CO_2C_2H_5$ with diazoacetic ester). 4-Diazophenoxides likewise coupled to give deeply colored quinone azines; **49** originated from 2,6-dibromo-4-diazophenoxide and diphenyldiazomethane.[61]

4-Nitrobenzenediazonium Ion as Dipolarophile

The reaction of diazonium salt **38** with diazomethane provided 1-(4-nitrophenyl)tetrazole (**50**, 12%) as well as **43**; with some reluctance from our side, an unusual coupling of **38** with N^β of diazomethane was assumed in 1955 to be the initiating step.[59]

In 1919 K. H. Meyer[62] asserted that 1,3-dienes are capable of azo coupling with arenediazonium ions; this theoretically significant finding became textbook knowledge. In 1975 Carlson, Sheppard, and Webster[63] revised the structures. Surprisingly, a Diels–Alder reaction of 2,3-dimethylbutadiene with the N≡N triple bond of the diazonium salt furnished **51**, which was deprotonated to the 1,6-dihydropyridazine derivative **52**.

In the light of this finding, our tetrazole **50** may spring from a 1,3-dipolar cycloaddition of diazomethane to **38** with subsequent aromatization by proton loss, as Franz Bronberger and I reasoned 30 years later. Azomethine ylides and thiocarbonyl ylides served as models for the novel dipolarophilic character of the aromatic diazonium ion. The mesoionic oxazolone **53** ("münchnone", page 102) rapidly interacted with 4-nitrobenzenediazonium fluoroborate; CO_2 extrusion from a bicyclic adduct produced the triazolium salt **54** (47%). Thiofluorenone-S-methylide (**55**), a thiocarbonyl ylide (page 110), was liberated from a precursor and combined *in situ* with the diazonium salt **38** to give **56** (49%) as deprotonated product.[64]

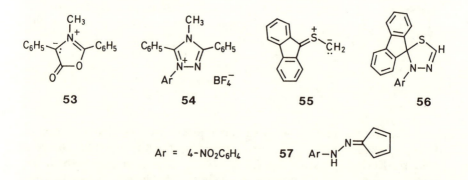

Ar = 4-NO₂C₆H₄   **57**

K. H. Meyer's 1919 claim[62] was not totally incorrect. Bredt's rule forbids deprotonation of the Diels–Alder adduct from cyclopentadiene and diazonium salt. Then azo coupling takes over. The reaction of **38** with an excess of cyclopentadiene at −50 °C allowed the isolation of derivatives of cyclopentadienone nitrophenylhydrazone (**57**), namely the mono- and bisadducts of cyclopentadiene to the C=C double bonds.[65]

Medium-Sized Ring Effects

For much of its "growing-up phase", organic chemistry was on the leash of natural product chemistry. Leopold Ruzicka (1926) recognized 15- and 17-membered rings in the animal odorants muscone and civetone. Synthetic methods soon gave access to large carbon rings, but yields were discouragingly small for rings with 9–12 members. "Medium-sized" rings (H. C. Brown, 1951), that is, those with 8–12 members, suffer not only from angle and conformational strain but also from a pressing of van der Waals radii, giving rise to a density maximum of cycloalkanes at the 10-membered ring.[66] Strain is diminished in large rings, which more readily assume normal bond angles and staggered conformations.

The minimum yield was less pronounced for the acyloin condensation of dicarboxylic esters as discovered independently by Vladimir Prelog and M. Stoll in 1947. Prelog's brilliant studies of reaction rates, equilibria, and transannular interactions of medium-sized ring compounds are regarded as classics today.[67]

A great deal of *general* knowledge has sprung from the investigation of reactions under *nongeneral* conditions. Small-ring chemistry proved prolific in this respect. Medium-sized rings had the charm of novelty in 1950, and I felt strongly tempted to apply them as model compounds.

Vladimir Prelog (right) talking with me at the Bürgenstock Conference in 1982. Prelog's erudition, competence, charm, and humor have won him many friends.

cis–trans Isomerism in Lactones and Cyclic Carbonates

Carbon chains favor a double bond in the *trans* configuration, whereas small and common rings tolerate only *cis* incorporation. The smallest *trans*-cycloalkene capable of storage is *trans*-cyclooctene with a strain energy of 17 kcal mol^{-1}.[68] Large cycloalkene rings accommodate *cis* and *trans* double bonds.

Because of the participation of zwitterionic resonance structures, carboxylic esters and carboxamides contain rotationally restricted bond systems. Their barriers to rotation are of intermediate size, between those of the carbon–carbon single and double bonds.

Fittig[69] discovered lactones and in 1881 called attention to a unique feature: the hydrolysis equilibrium of six-membered lactones is established in cold aqueous solution. In contrast to open-chain esters, lactones can be titrated by 0.1 N NaOH. In the 1940s the *alkylating* capacity of β-lactones was ascribed to the ring size, but as early as 1886 Wislicenus[70] observed an analogous conversion of five-membered lactones by potassium cyanide into 4-cyanocarboxylic acids. Analogous? A humorist defined analogy as a linking of a new unknown with an old unknown.

Resonance makes the carboxylic ester group quasiplanar. Our reasoning that the exceptional reactivity of lactones originates from the unfavorable *cis* conformation of the ester group found ample support in the comparison of physical and chemical properties of 5- to 16-membered lactones **58**,[71] carried out by Heinz Ott (Ph.D. 1955).

The discontinuity in the *dipole moments* of the homologous series **58** (Table II) allows us to locate the turnover from *cis*- to *trans*-ester group. A dissection into partial moments is helpful. The C=O bond moment and the one stemming from the zwitterionic contribution may be combined the *carbonyl* moment; the polarity of the two C–O bonds and the lone-pair moment add up to the *ether* moment. The vector sum of carbonyl (~3.3 D) and ether (~1.6 D) moments simulates the dipole

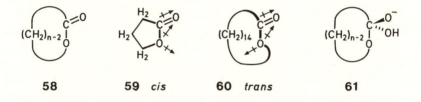

58   **59** *cis*   **60** *trans*   **61**

Table II. Dipole Moments, Boiling Points, Rate Constants of Alkaline Hydrolysis, and Hydrolysis Enthalpies of Lactones

n (58)	Dipole Moment[a] (Debye)	Boiling Point[b] (°C)	Rate Constant[c] ($M^{-1} s^{-1}$)	Hydrolysis Enthalpy[d] (kcal/mol)
5	4.09	80	1,480	−2.1
6	4.22	97	55,000	−3.1
7	4.45	106	2,550	−4.8
8	3.70	80	3,530	e
9	2.25	72	116	e
10	2.01	86	0.22	e
11	1.88	100	0.55	e
12	1.86	116	3.3	−1.0
13	1.86	130	6.0	−1.1
14	1.86	169	6.5	2.0
16	1.86	169	6.5	1.6
Butyl caproate	1.79	83	8.4	1.8

[a]In benzene, 25 °C.
[b]At 1.3 kPa (10 torr).
[c]$10^4 k_2$, in 60% dioxane, 0 °C.
[d]In 60% dioxane.
[e]No measurement.

moment values of *cis*-lactones (4.09 D for butanolide, **59**) and *trans*-lactones like exaltolide (**60**, 1.86 D).

The value of 1.86 D for **58** (n = 12–16) agrees well with open-chain aliphatic esters (1.79 D) for which electron diffraction established the *trans* structure.[72] Hence, lactones switch from *cis* to *trans* configuration as soon as the diminishing ring strain permits. Intermediate values for **58** (n = 8–10) suggest *cis–trans* mixtures (Table II).[71]

Because of dipole association, the *boiling points* of *cis*-lactones are higher than those of *trans*-lactones by ~70 °C at 10 torr, based on CH_2 increments (14 °C). Butyrolactone ($C_4H_6O_2$) boils at 206 °C, and ethyl acetate ($C_4H_8O_2$) at 77 °C, both at normal pressure. Within the lactone series, hexanolide boils higher than decanolide (Table II).

The *rate constants of alkaline hydrolysis*, k_2, respond to the conformational change *cis* → *trans* with a decrease of 4–5 orders of magnitude (Table II). The k_2 values of *trans*-lactones **58** (n = 10 and 11) are lower than those of the higher homologues; the latter (n = 12–16) approach the k_2 of open-chain esters. *trans*-Lactones are favored over *cis*- by ~4 kcal mol^{-1} under the simplifying assumption that the energy difference is wiped out in the transition state of hydroxide addition leading to **61**. Like hydrolysis, nucleophilic attack at the ω-position of *cis*-lactones profits from this energy difference, the carboxylate anion being the leaving group.[71]

In the *cis–trans* transition region, the dipole moment and k_2 of **58** (n = 8) illustrate *cis* predominance, but its boiling point is ~40 °C too low for a *cis*-lactone; it vaporizes via a small *trans*-lactone share. The dipole moment of **58** (n = 9) is only 0.24 D higher than that of **58** (n = 10), but its hydrolysis is 500 times faster. This behavior indicates a mixture of *trans*- with some *cis*-lactone (hydrolysis via the latter).

Since our 1959 paper was published,[71] a small concentration of the *cis* conformation was established by IR, UV, and ^{13}C NMR spectroscopy in open-chain carboxylic esters. Blom and Günthard[73a] froze out the equilibrium mixture of *trans*- + *cis*-ester from thermal molecular beams; the analysis of the C=O stretching modes gave a reliable value only for methyl formate: $\Delta H°(cis–trans)$ = 4.8 ± 0.2 kcal mol^{-1}. Ab initio calculations with large basis set by Wiberg and Laidig[73b] provided the high values of 6.0 and 9.3 kcal mol^{-1} for $\Delta E(cis–trans)$ of methyl formate and methyl acetate.

In 1963 Roland Krischke in our laboratory measured the *hydrolysis enthalpies* (ΔH_{liq}) of *cis*- and *trans*-lactones (dioxane:water 3:2, Table II). The data remained unpublished because the energy difference of the *cis*- and *trans*-ester group is intertwined with the conformational and Baeyer strain. The hydrolysis enthalpies (ethanol:water 3:2) of **58** (n = 5–14), as measured by Wiberg and Waldron[73c] in 1991, correspond well with our data. The strain energies derived did not allow a clear correlation with the conformation of the ester group. In contrast to cyclic ketones and ethers, the strain energy of δ-valerolactone (**58**, $n = 6$) exceeds that of **58** ($n = 5$), a hitherto unexplained phenomenon; heats of combustion likewise marked **58** ($n = 6$) as exceptional.[73d]

Two C–O bonds with partial double-bond character occur in dialkyl carbonates, allowing for three conformations in *cyclic carbonates*. The partial dipole moments coming from the zwitterionic structures are codirectional with the C=O bond moment and add to it (Table III). The vector sum of the two ether moments, in turn, is likewise coaxial with the carbonyl moment, increasing it in the *cis,cis* conformation and opposing it in the *trans,trans* arrangement.

Table III. Physical Properties of Alkaline Hydrolysis in 60% Dioxane at 0 °C for Cyclic Carbonates and Diethyl Carbonate

| | **62** | **63** | **64** | |
| | *cis,cis* | *trans,cis* | *trans,trans* | *trans,trans* |

Property	62	63	64	*Diethyl Carbonate*
Dipole moment (Debye, benzene, 25 °C)	5.31	3.59	1.20	0.99
Boiling point (°C, 0.3 torr)	90	69	110	~–25
Rate constant, $10^3 k_2$ (M^{-1} s^{-1})	3.070	3.74	0.43	1.61

We applied the set of structural probes to cyclic carbonates from the 5- to the 19-membered ring (Table III).[74] Roland Krischke measured the dipole moments and the hydrolysis enthalpies, whereas Edmund Stelter described the rates of alkaline hydrolysis (1965). The formidable level of the dipole moment of *cis,cis*-carbonates, 5.31 D for **62**, drops to 1.20 D for the 16-membered *trans,trans*-conformation of **64**. On their way, the μ values pass through an intermediate level, such as 3.59 D for the 11-membered *trans,cis*-carbonate **63**.

Two discontinuities in the plot of boiling points versus ring size lend credence to the existence of three defined conformations. The bp$_{0.3}$ of *trans,trans*-carbonates with a 12–19-membered ring increases linearly by 11 °C per methylene unit; *trans,cis*-carbonates boil "too high" by 25 °C, whereas the *cis,cis*-carbonate **62** shows $bp_{0.3}$ at 90 °C instead of at −10 °C as extrapolated from the *trans,trans* function!

The rates of alkaline hydrolysis decrease by 10^4 for *cis,cis-* > *trans,cis-* > *trans,trans*-carbonates, the borders being blurred by the presence of multiple conformations. The first step leading to an open-chain monoalkyl carbonate is fast for *cis,cis*-carbonates (1:1 stoichiometry), whereas k values of the first and second hydrolysis step are of a similar magnitude for *trans,trans*-carbonates (consecutive system).[74]

Conformation of Lactams and *N*-Nitrosolactams

Two planar conformations are anticipated for *N*-monosubstituted amides. X-ray analyses of acylamino acids and glycylglycine disclosed planar *trans*-amide groups **65**. The *trans* preference was also postulated for the state in solution, a prerequisite of L. Pauling's illustrious helix model of protein structure.[75,76]

Several authors assigned the *trans* structure **65** by comparing *dipole moments* with values calculated from bond moments for both conformations.[77,78] In reality, dipole moments are not indicative here. We measured those of lactams for $n = 5$–19 (benzene, 25 °C): μ for **66** grows from 3.55 D ($n = 5$) to 3.88 D ($n = 7$); it slightly diminishes to 3.67 D for **67** ($n = 17, 19$).[79] The dipole moments of cycloalkanones are smaller, but show the same trend with growing ring size.[80]

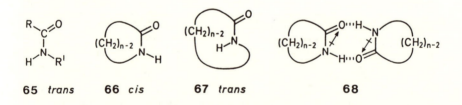

65 *trans*   **66** *cis*   **67** *trans*   **68**

However, the dependence of the *molar polarization* P_2 on the mole fraction γ_2 distinctly demonstrated the *cis* structure **66** ($n = 5$–9) and *trans* conformation **67** ($n = 10$–19) (Figure 1); all the dielectric measurements were done by Helmut Walz. The formation of cyclic dimers **68** from *cis*-lactams is accompanied by a decrease of polarity, whereas the chain association of *trans*-lactams goes along with rising polarity because of the phenomenon of induced moments.[79]

Numerical evaluation of the functions in Figure 1 afforded free energies $\Delta G = -2.8$ to -3.4 kcal mol^{-1} for the dimerization of *cis*-lactams **66** ($n = 5$–8) and dipole moments $\mu = 2.2$–2.6 D for the dimers.

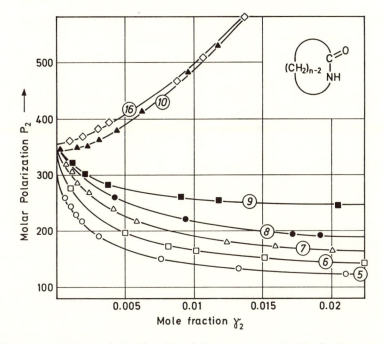

Figure 1. *Molar polarization P_2 of lactams, n = 5–16, in benzene at 25 °C as a function of mole fraction γ_2; ring size in circles. (Reproduced with permission from reference 79. Copyright 1956 VCH Verlagsgesellschaft mbH.)*

In the chain associates, the units of the *trans*-lactam are linked by only one hydrogen bond; the association is still incomplete at $\gamma_2 = 0.015$.

No less significant is the *amide II band* in the infrared spectrum, which is ascribed to a combination of C–N stretching and N–H bending mode. U. Schiedt, my colleague in Tübingen and Munich, noticed[81] in 1954 that this band, present in all open-chain amides **65**, was missing in the *cis*-lactams **66** (n = 5–9). In the chloroform solution of **67** (n = 10–19), the amide II band appears at 1515–1520 cm^{-1}.[82] N–H frequencies likewise allowed structural assignments of *N*-alkylamides.[83]

Caprylolactam (n = 9) is the borderline case. A weak amide II signal suggests that ~12% of *cis*-lactam **66** is in equilibrium with *trans*-form **67** in chloroform, whereas the crystal contains only **67**.[82] In 1975, an X-ray analysis of **67** (n = 9) by Dunitz and Winkler[84] confirmed our *trans* assignment of 1957; an out-of-plane deformation and 17° torsion at the C–N bond indicate ring strain.

How did the problem of carboxamide configuration arise? It was tied to our study of the isomerization of *N*-nitrosoacylamides to *trans*-diazoesters (page 18). This acyl shift showed a dramatic rate increase by

10^9 in the series of N-nitrosolactams **69** (n = 5–9) and a decrease by factors of 130 and 370 for **70** (n = 10, 11) respectively.[27]

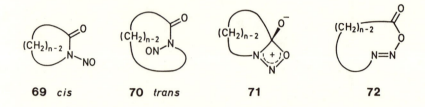

69 *cis*   **70** *trans*   **71**   **72**

The steep rate maximum suggested two opposing forces. One force is Baeyer strain in the transition state leading to **71** and strain in the diazoester **72**, which harbors a *trans* N=N double bond. The second factor is the jump from the enforced *cis*-conformation **69** (n = 5–9) to the favored *trans*-form **70** ($n \geq 10$). This phenomenon focused our interest on the *cis,trans* isomerism of the more important carboxamide and ester groups.

Intramolecular Friedel–Crafts Acylation

K. Ziegler's synthesis of medium-sized and large rings by base-catalyzed cyclization of dinitriles made use of *high dilution* to suppress intermolecular condensation.[85] Will the "Ruggli–Ziegler principle" also give access to medium-sized ring ketones by intramolecular Friedel–Crafts acylation? Cyclophane chemistry evolved in the 1950s and the expected products might offer a playground.

The following conditions were found effective: 0.1 mol of ω-phenylalkanoic acid chloride **73** in 2 L of CS_2 was introduced into 4 L of boiling CS_2 that contained either the poorly soluble $AlCl_3$ or the more soluble $AlBr_3$. The reflux was regulated in such a way that the acid chloride entered the flask in 300 L of solvent over 120 h.[86]

1,2-Benzocyclooctene-3-one (**75**, n = 8) was obtained in 76% yield from **73** (n = 5); a scant yield of **75** (n = 9) resulted with $AlBr_3$ as catalyst.[86–88] The higher homologues (**73**, $n \geq 8$) chose ring closure at the 4-position. The yields of paracyclophanones **77** (n = 13–18) rose with increasing ring size.[86,89] The ω-(4-tolyl)alkanoic acid chlorides **74** (n = 6,7) provided the o-ketones **76** (n = 8, 9); acylation of **74** (n = 9–16) by $AlBr_3$ proceeded in the methyl-activated 3-position, affording meta-cyclophanones **78** (n = 12–19).[90]

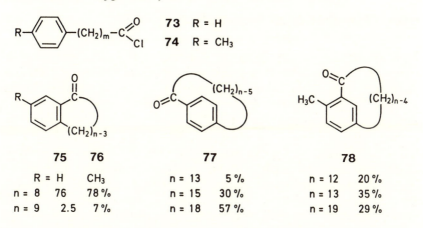

75	76	77	78
R = H	CH$_3$	n = 13 5 %	n = 12 20 %
n = 8 76	78 %	n = 15 30 %	n = 13 35 %
n = 9 2.5	7 %	n = 18 57 %	n = 19 29 %

The 1- and 2-naphthylalkanoic acid chlorides studied by Ulrich Rietz (Ph.D. 1955) preferred the 1,7- to the 1,4-annellation (e.g., furnishing **79**, $n = 11$–14).[91,92] The 7-position of ω-(6-tetralyl)alkanoic acid chlorides is activated by o,p-dialkyl. Much to our delight, cyclization gave rise to 6,7-annellation exclusively, thus affording the series of cyclanones **80** ($n = 6$–22).[95] This was a stroke of luck, and the relatively high yields of medium-sized rings **80** were a fringe benefit. This long homologous series provided Viktor Trescher[93] and Harald Oertel[94] with ideal models for the conformational studies to be described.

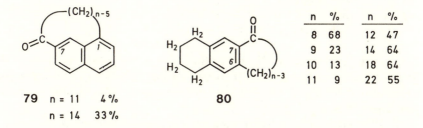

n	%	n	%
8	68	12	47
9	23	14	64
10	13	18	64
11	9	22	55

79   n = 11   4 %
        n = 14   33 %

80

Chemists are pleased when their newly developed methods find wide acceptance. The echo to our high-dilution cyclization, elaborated by Walter Rapp and Ivar Ugi in their Ph.D. theses, was not overwhelming. Obviously, fellow chemists got cold feet over the idea of an arsenal of big flasks, each containing 5–6 L of boiling carbon disulfide, although we never had a fire. In 1963 Gol'dfarb et al.[96] cyclized ω-(2-thienyl)alkanoic acid chlorides in high dilution in chloroform by partially hydrolyzed $AlCl_3 \cdot O(C_2H_5)_2$. Acylation in the 5-position gave the 12–16-membered cyclic ketones in 54–64% yield, whereas the minimum in the medium-sized range was pronounced.

Steric Hindrance of Resonance

Light absorption sensitively signals the balance of forces controlling the structure. The aesthetic first evidence for steric hindrance of resonance came from the bright colors of the crystalline 2,4-dinitrophenylhydrazones 81: $n = 6$, garnet red; $n = 7$, light scarlet; $n = 8$, yellow-orange, like the cyclohexanone derivative.[87] In 2-*tert*-butylacetophenone the carbonyl group is *pushed out* of coplanarity with the benzene ring by the bulky substituent. On going from $n = 6$ to 9, the hydrazone group in 81 and the carbonyl group in 75 are *pulled out* of the aromatic plane by the strained polymethylene chain. According to space-filling models, the CO group in 75 ($n = 9$) should be oriented orthogonally to the aromatic plane. However, the loss of conjugation energy limits the twisting. Considering the decreasing oscillator strength of the $\pi \to \pi^*$ absorption, we estimated that torsion angles amount to 29°–44° for 75 ($n = 7$–9), based on 1-tetralone (75, $n = 6$) being planar.[88]

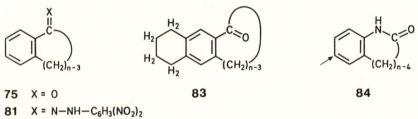

75 X = O   **83**   **84**
81 X = N—NH—C$_6$H$_3$(NO$_2$)$_2$
82 X = N—OH

The UV spectra illustrate that the conjugative coupling of aromatic oximes 82 is weaker than that of ketones 75. The strong $\pi \to \pi^*$ band of 1-tetralone oxime (82, $n = 6$) completely disappeared in both *syn*- and *anti*-82 ($n = 8$).[97]

The molecular model of paracyclophanone 77 ($n = 13$) suggested orthogonality of carbonyl and aromatic plane and a restoration of planarity for $n = 15$. The $\pi \to \pi^*$ transitions of 77 ($n = 13$–15) revealed 45%, 80%, and 84% of the extinction coefficient of 4-methylacetophenone.[88] An enlightening message came from the UV spectra of the tetralinocyclenones 80 ($n = 5$–22). The minimum molar extinction value ϵ was observed for $n = 10$ with a torsion angle of 35°; with growing ring size, ϵ leaps up for $n = 14$, but has not fully recovered for $n = 22$. Molecular models suggest the O-inside conformation 83 for large rings.[95]

In the 1,2-benzolactams 84, the aniline-type resonance is weaker than the competing carboxamide mesomerism. The ϵ value of the long-wave absorption of **84** ($n = 6$) decreases with rising n, associated with a hypsochromic shift; the vanishing of the band in $n = 9$ convinced Heinz Brade and me that conjugative coupling of C_{ar}–N is "soft".[98]

Steric hindrance of resonance controls reactivity, too. Conjugation is very strong in benzyl-type cations. The ethanolysis rates of 1,2-benzo-3-chlorocyclenes **85** at 40 °C diminishes 1000-fold from $n = 5$ to 8, reflecting growing torsion;[97] a different dependence on ring size for the solvolysis of 1-methylcycloalkyl chlorides[99] indicates that decreasing benzyl-type resonance is the major contributor.

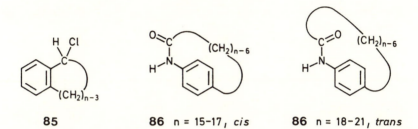

85   **86** n = 15–17, *cis*   **86** n = 18–21, *trans*

The bromine substitution of 1,2-benzolactams **84** in the p-position is rather selective, as a 900,000-fold decrease in k_2 (HOAc, 20 °C) in the range from $n = 5$ to 9 attests. With the loss of the competing aniline resonance, the basicity of the amide group increases by 1.4 pK units.[98]

In the 1,4-benzolactam **86** ($n = 15$), the amide group is twisted by ~40° versus the benzene ring. The oscillator strengths of the UV absorption diagnose a stepwise return to coplanarity with growing ring size; the band of **86** ($n = 21$) is identical with that of aceto-p-toluidide. The rate constant of bromine substitution ortho to the amide function increases 1700-fold in going from $n = 15$ to 21. Ivar Ugi found that the switching from *cis*- to *trans*-amide group is superimposed; the diagnostic amide II absorption (KBr disc) is missing in **86** ($n = 15$–17), but occurs in $n = 18$–21.[100]

Bridged Carbocations in Solvolysis

In the 1950s, Saul Winstein's studies on the participation of neighboring groups in ionizations (page 23) received the attention they deserved and stimulated a wealth of research. The neighboring group's assistance lowers the energy level of the cationic intermediate. The participation of phenyl in solvolysis was discovered by Don Cram[101] in 1949; the stereochemistry observed in the acetolysis of *erythro-* and *threo*-2-phenyl-3-butyl tosylate and related systems suggested a phenonium ion as intermediate. The steric course in the formolysis (70% ee of retention) and acetolysis (30% ee of inversion) of optically active 1-phenyl-2-propyl tosylate revealed a rather weak phenyl assistance. The solvolysis rates were lower than those of isopropyl tosylate.[102]

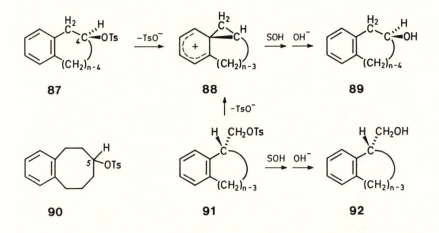

87   88   89

The closure of the three-membered phenonium ring freezes internal rotations (i.e., the decrease of entropy reduces the importance of phenyl participation). A rotationally restricted cyclic system should be free of this adverse effect. The 1,2-benzocycloalkenyl 4-tosylates **87** offer beautiful models for intracyclic phenonium formation, and the length of the polymethylene chain allows variation of ring strain in the bridged **88**. The rate constants of formolysis of **87**, measured by Erich Rauenbusch, indeed fit the formation of the phenonium ion **88** (k_{rel} based on $n = 6$):[103]

87, n	6	7	8	9
k_{rel} (35 °C)	≡1	5.6	82	27

In space-filling models, the four CH_2 groups in **88** ($n = 7$) formed from **87** ($n = 8$) allow a strain-free bridging of the *o*-positions in the spiro system, whereas the five CH_2 groups in **88** ($n = 8$) cause a slight "overwinding". The formolysis rate constant of 1,2-benzo-5-cyclooctenyl tosylate **90** was slower by a factor of 1000 than that of **87** ($n = 8$), thus demonstrating the specific effect of β-aryl.

A bridge of two CH_2 groups is too short to allow the formation of the phenonium ion (**88**, $n = 5$) from β-tetralyl tosylate (**87**, $n = 6$). Formolysis and acetolysis of **87** ($n = 6$) yielded after alkaline hydrolysis β- and α-tetralol as well as 1,2- and 1,4-dihydronaphthalene, classic solvolysis products of *sec*-alkyl tosylates. In contrast, after the formolysis of **87** ($n = 7$ or 8), 96% of **89** ($n = 7$ or 8) was isolated and no olefin was detected. *Formolysis* of optically pure (−)-**87** ($n = 7$) proceeded with 100% retention as result of double inversion. The first inversion occurred in the formation of **88** ($n = 6$) by phenyl assistance, the second in the opening of the three-membered ring by solvent attack. In the *acetolysis* of (−)-**87** ($n = 7$), the retention rate was diminished to 89% ee

Ten full papers were the offspring of Günther Seidl's Ph.D. thesis (1958) and two postdoctoral years in Munich. In 1960 Seidl joined the pharmaceutical synthesis at the Hoechst AG, Frankfurt; after several interludes, partly in the United States, Seidl directed the pharmaceutical research of the company 1976–1988. This was the exciting period when the broad empirical screening of new classes of compounds was replaced by a more biological approach based on receptor interactions, enzyme inhibition, and physiological control mechanisms. (This photograph was taken in 1958.)

and the product contained 8% olefin.[104] Thus, all the data competently elaborated by Günther Seidl underline the dominant role of the bridged ion **88**.

The branched-chain primary tosylates **91** ($n = 6$ or 7) quantitatively afforded the ring-enlarged alcohols **89** ($n = 7$ or 8) via **88** on *formolysis* (always followed by ester hydrolysis); the retention reached >93% ee when optically active **91** ($n = 6$) was employed. The formation of ~20% of the primary alcohols **92** ($n = 6$ or 7) along with **89** ($n = 7$ or 8) in the *hydrolysis* (70% dioxane, 90 °C) of **91** ($n = 6$ or 7) pointed to a competing pathway of the S_N2 type.[105]

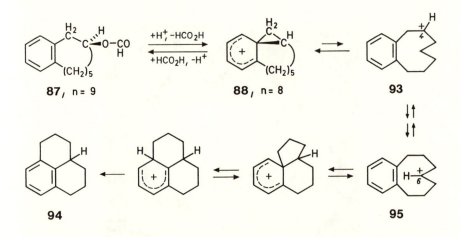

In the solvolysis of tosylates in *unbuffered* formic acid, the released toluenesulfonic acid can generate the carbocation again and again, and the final outcome is sometimes hard to predict. I still remember Günther Seidl's and my amazement when prolonged treatment of tosylate **87** ($n = 9$) or **91** ($n = 8$) in formic acid at 65 °C produced 65–80% of tetrahydroperinaphthane (**94**). The phenonium ion **88** ($n = 8$) is reversibly formed and the slow concomitant pathway via **93** becomes dominant. Two successive 1,2 hydride shifts lead to **95**, and transannular alkylation finally affords **94**.[106]

Aryl Migration in Beckmann Rearrangement

The term "carbonium salt" originated with Adolf von Baeyer. In 1902 he[107] compared triphenylmethyl sulfate with aluminum sulfate because both produce insoluble hydroxides with aqueous alkali. Twenty years later, Hans Meerwein[108] recognized the importance of carbocationic intermediates in molecular rearrangements of the camphene hydrochloride type, later named after Wagner and Meerwein.

Like Wagner–Meerwein rearrangements, the conversion of ketoximes into carboxamides (named after E. Beckmann) is induced by acids or other electrophilic reagents. As a sequel to Hantzsch's discovery of the *syn–anti* isomerism of oximes in 1890, the Beckmann rearrangement was often applied to the stereochemical assignment. This application was based on the naive assumption that alkyl or aryl moves to nitrogen on the same side that harbors the hydroxyl in the reactant. J. Meisenheimer's discovery of the obligatory *trans* migration[109,110] was a milestone in the development of mechanistic concepts for sextet rearrangements. Heterolysis of the N–O bond and migration of R occur *simultaneously* in the protonated oxime **96**. Thus, the nitrenium ion (i.e., a sextet structure) is shunned in favor of the nitrilium ion **97**, a decent octet structure.

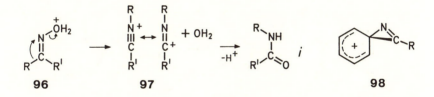

The migratory aptitude of aryl exceeds that of alkyl in molecular rearrangements. The activation energy of phenyl migration to the adjacent C atom is lowered by the cyclohexadienyl resonance that the phenonium cation has in common with σ-complexes in the electrophilic substitution of benzene. Is it conceivable that phenyl migration to adjacent nitrogen in the Beckmann rearrangement likewise passes a bridged structure? In the 1950s we conjectured the spiro-azirine **98**, more strained than the analogous phenonium ion, to be an intermediate in aryl migration. Today the subject of a rich chemistry,[111] 2*H*-azirines were then unknown except for their supposed occurrence (based on one example isolated) in the alkoxide-induced Neber rearrangement of oxime tosylates to give α-aminoketones.[112]

Our spectrophotometric rate measurements concerned the Chapman modification.[113] Josef Witte (1927–1988; Ph.D. 1957) measured the rate constants k_1 for the thermal rearrangement of substituted acetophenone oxime 2,4,6-trinitrophenyl ethers **99** to the imidoyl esters **100**, which, in turn, suffer a quick trinitrophenyl shift to yield the N-trinitrophenyl carboxamides **101**. In accordance with the formation of a type-**98** intermediate, **99**, S = 4-CH$_3$O reacted 700,000 times faster than S

= 4-NO$_2$ (1,4-dichlorobutane, 70 °C). The log k_1 of 12 *m*- and *p*-substituted **99** fulfilled the Hammett equation with $\rho = -4.1$.[114] The rate constants for **99**, S = 4-CH$_3$O and 4-C$_6$H$_5$, were found "too high"; H. C. Brown's σ^+ scale (1958) was not yet defined. Pearson et al.[115] dealt with the Beckmann rearrangement of substituted acetophenone oximes in concentrated sulfuric acid and came to similar mechanistic conclusions, although the rate-determining step admittedly was not clear.[116]

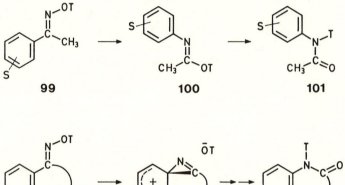

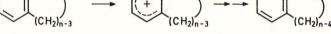

T = 2,4,6-Trinitrophenyl

As a crucial test, we studied the intracyclic variant. Molecular models of **103** illustrate that at least five CH$_2$ groups ($n = 8$) are required to keep the bridge strain-free. The rate constants of the Chapman rearrangement of the *anti*-oxime trinitrophenyl ethers

102, $n=$	5	6	7	8
$10^6\,k_1$ (s^{-1}, 70 °C)	0	<0.02	1870	429,000

disclose a >20 millionfold rate increase from $n = 6$ to 8 as a result of two cooperative phenomena: diminution of ring strain in the bridged intermediate **103** and loss of conjugation between the aromatic ring and the C=N double bond in **102**.[117] The aryl migration in **102** ($n = 8$) was 175,000 times faster than alkyl migration in the stereoisomeric *syn*-oxime trinitrophenyl ether.

Thus, the reason for the high migratory aptitude of aryl groups in sextet rearrangements appears to be clarified. All kinetic evidence points to bridged structures, that is, to an involvement of the migrating phenyl in an electrophilic substitution.

Electrophilic Azo Compounds and Azomethine Imines

Aliphatic azo compounds vegetate in the shadow of their aromatic brethren; aromatic azo dyes enrich the world by bright colors. The unusual reactivity spectrum, rather than the optical spectrum, makes up the charm of aliphatic azo compounds, especially those with electron-attracting substituents.

Azodicarboxylic Ester, an Acrobat of Reactivity

The record of Theodor Curtius's discoveries in nitrogen chemistry is stunning: diazoacetic ester (1883), hydrazine (1887), hydrazoic acid (1890), and azodicarboxylic ester (1894).[118] Otto Diels systematically studied nucleophilic additions of bases HB to the N=N bond of diethyl azodicarboxylate (105) in the 1920s. Acid-catalyzed electrophilic aromatic substitution by 105 provided N-arylhydrazine-N,N'-dicarboxylic esters, as observed by Curtius's former student, R. Stollé.[119] The adduct 104, formed from 105 and cyclopentadiene, constituted the first clarified example of a Diels–Alder reaction (1925).[120] Three years later the generality of this synthetic principle was recognized.[121]

Analogies often motivate research. I wondered whether the behavior of azodicarboxylic ester toward alkylbenzenes would parallel that of bromine. Nuclear bromination is an electrophilic substitution, whereas the introduction into the benzylic position takes place via a radical chain. Kharasch, Mayo et al.[122] (1938) had established the nature of the side-chain bromination.

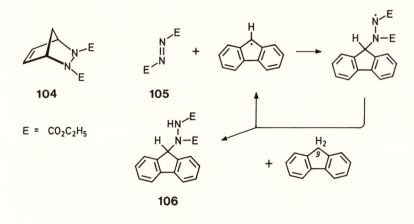

104   **105**

E = CO₂C₂H₅

106

In 1951, still at Tübingen, we encountered a suitable model in fluorene + diethyl azodicarboxylate. The addition of the 9-CH of fluorene to the N=N of **105** afforded **106** and was indeed accelerated by diacyl peroxides or azoisobutyronitrile (AIBN). The inhibition by quinones was exceeded by 2-*tert*-butyl-4-methoxyphenol and, remarkably, acrylonitrile.[123] Anton Cadus (Ph.D. thesis 1952) was analytically minded; he was not satisfied until the mother liquors were exhausted. We depended on the isolation of **106** because physical methods were not yet available. The ethoxycarbonyl radical formed by slow thermal homolysis of **105** is a likely candidate for the spontaneous initiation of the radical chain.

The substitution of alkylbenzenes by **105** can be directed as is that by bromine. Under catalysis by concentrated sulfuric acid, *p*-xylene was converted by **105** into **107** and the bisadduct **108**; the donor group X in **107** facilitates the second electrophilic substitution. On the other hand, the thermal reaction of **105** with *p*-xylene furnished **109** and some **110**.[124]

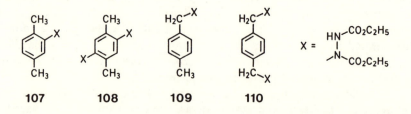

107   **108**   **109**   **110**

Ketones are attacked in the α-position by **105**. The rate of conversion of cyclohexanone into adduct **111** was increased by diacyl peroxides and lowered by benzoquinone. Franz Jakob (Ph.D. thesis

1954), a very efficient co-worker, found catalysis by potassium acetate or concentrated sulfuric acid even more effective. Supposedly, it involves the addition of the enolate anion or the enol, respectively, to the N=N bond of **105**.[124]

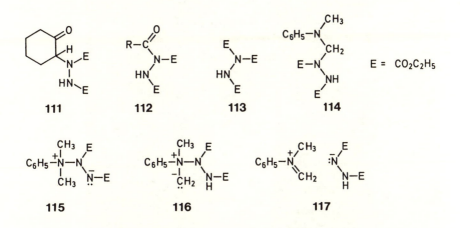

According to Alder and Noble,[125] aldehydes add to **105** at room temperature, affording N-acylhydrazodicarboxylic esters **112**; the analogy with the autoxidation of aldehydes furnishing peracids was emphasized. Indeed, we observed a strong inhibition of the interaction of acetaldehyde with **105** by p-benzoquinone. At 100 °C, ethyl formate added to **105** and the rate of formation of **113** shot up in the presence of AIBN.[124]

On my first trip to the United States in 1955, I lectured at the University of Chicago on diazo and azo chemistry. Afterward, Morris Kharasch showed me his card file with unpublished work. Radical chain additions of azodicarboxylic ester to the aromatic side chain were familiar to him. The leading radical chemist gladly accepted the news that a young German scientist had entered his research area. During the evening in his home, Kharasch's personality impressed me as much as reading his papers did before. His premature death was a tremendous loss for radical chemistry.

In 1922 Diels[126] described a 1:1 adduct of **105** and N,N-dimethylaniline, for which structure **114** was proposed. The acid hydrolysis of **114** giving N-methylaniline, formaldehyde, and hydrazodicarboxylic ester characterizes azo ester **105** as a demethylating reagent. Our kinetic measurements revealed a second-order conversion, not responding to radical inhibitors. The rate constant rose with increasing solvent polarity; 3-nitro-N,N-dimethylaniline reacted 25 times slower than N,N-dimethylaniline, whereas the 4-nitro compound was inert.[124]

Morris B. Kharasch (1895–1957) at the University of Chicago in 1955. His ima-
gination and his experimental mastery made Kharasch one of the great pioneers of
radical chemistry.

We postulated the zwitterionic intermediates **115** and **116**; Kenner and
Stedman[127] (1952) came to the same conclusion. Today I would rather
replace the ylide **116** by the iminium ion pair **117**.

In 1943 Alder et al.[128] described allylic substitutions of olefins by
105 as well as by maleic anhydride, stressing the formal resemblance to
autoxidation. The authors were astounded by the shift of the double
bond in the case of allylbenzene; it was "no genuine substitution" and
was considered exceptional. In 1949 Alder et al.[129] became aware that
the shift of the double bond was the rule in the "indirect substitutive
addition" of alkenes by maleic anhydride. The scheme corresponds to
an *ene reaction*,[130] and Arnold and Dowdall[131] were the first to postulate
a cyclic transition state for these.

In fact, two mechanisms contribute to the allylic substitution by
the azo ester. In refluxing cyclohexene, adduct **118** was formed. We
noticed a sixfold increase in the conversion rate when dibenzoyl peroxide or AIBN was added. However, inhibitors cut the thermal reaction
only to half the rate.[124]

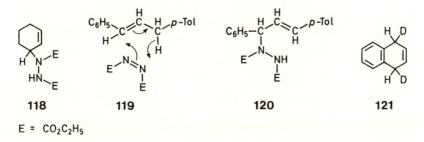

118   **119**   **120**   **121**

E = CO₂C₂H₅

Cyclohexene offered no probe for the double-bond shift. We chose 1-phenyl-3-*p*-tolylpropene as the second model. The exclusive formation of **120** in the interaction with **105** demonstrated the obligatory shift of the double bond. This shift was likewise shown for the propene with exchanged aryl groups. The rate did not respond to radical initiators or inhibitors; this behavior is consistent with a one-step cyclic process via **119**.[132]

The reaction of **105** with 1,2- and 1,4-dihydronaphthalene similarly proceeded with complete migration of the double bond. Again the rate was not influenced by initiators or inhibitors. An intramolecular kinetic isotope effect, $k_H/k_D = 3.7$ at 60 °C, was measured by applying the dideuterio derivative **121**.[132] Stretching and bending modes were involved, and the size of the isotope effect appeared reasonable for the cyclic process. This kinetic isotope effect, measured by Hanns Pohl, was probably the first one applied to an ene reaction.

In a reinvestigation in 1964, Thaler and Franzus [133] confirmed our observation of a radical chain for the initiated reactions of **105** with cyclopentene and cyclohexene. Amusingly, these cycloalkenes are exceptions. For open-chain alkenes, the substitution was not induced by dibenzoyl peroxide. The double-bond shift is the rule, and Franzus[134] referred to the same cyclic transition state we had proposed for 1,3-diarylpropenes.

In the system 1,3-cyclohexadiene + **105**, the ene reaction furnishing **122** represses the Diels–Alder addition.[134] However, irradiation afforded 87% Diels–Alder adduct **123**.[135] According to Schenck et al.,[136] a photostationary equilibrium of **105** with the *cis* isomer **124** is established, and **124** is the more potent dienophile.

The kaleidoscope of azo ester reactivity shows more colors in the interaction with donor olefins. Alder and Niklas[137] described β-substitution of α-methoxystyrene giving rise to **125**. Koerner von Gustorf et al.[138] observed 1,2-diazetidines from simple vinyl ethers and **105** (e.g., **126**), probably formed via a zwitterionic intermediate. The addition of **105** to vinyl acetate afforded the (2+4) cycloadduct **127**; 1,2-dimethoxyethylene and indene behaved analogously. The authors preferred a concerted addition to the two-step pathway via zwitterion.

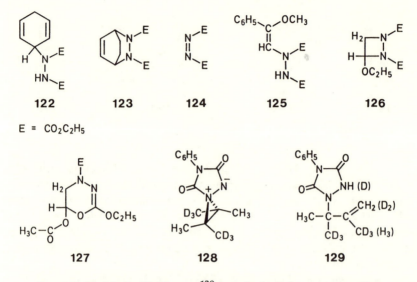

E = CO$_2$C$_2$H$_5$

In 1969 H. M. R. Hoffmann[130] concluded that ene reactions may occur via a broad spectrum of transition states. A brand-new idea sprang up in 1980. 1,2,4-Triazoline-3,5-diones are cyclic imides derived from 124; they match azo ester in the variety of reactions, showing even higher rate constants. Guided by the analogy with Br$_2$ and singlet O$_2$ furnishing bromonium zwitterions or the perepoxide with alkenes, respectively, Seymour and Greene[139] postulated an initial 1,1 cycloaddition of triazolinediones to the C=C double bond. The aziridinium imide of type 128 rearranges to the ene product 129. An ingenious experiment supported the spiro structure 128: only the *trans*-form of the D-labeled substrate disclosed an intramolecular isotope effect of 3.8 in the product. In 1985 Nelsen and Kapp[140] gained NMR evidence for such a spiro intermediate from triazolinedione and biadamantylidene, here on a pathway to a diazetidine. The resemblance to singlet O$_2$ reactions (i.e., the formation of 1,2-dioxetanes and ene products via perepoxide) is striking.

The scenario of ene reactions and (2+2) cycloadditions of azo ester via intermediates of type 128 may well be similar and a renewed study could be rewarding. The plethora of reaction modes shown by azo ester and its derivatives is spellbinding. What will be the next surprise?

Electrophilic Azo Compounds and Diazoalkanes

In Curtius's laboratory, E. Müller[141] mixed diethyl azodicarboxylate with ethyl diazoacetate; the diaziridine structure 132 was ascribed to the product. In the 1960s **132** was replaced by the 1,3,4-oxadiazoline formula **131**;[142,143] above 100 °C, the hydrazone-*N,N*-dicarboxylic ester **133** was formed.

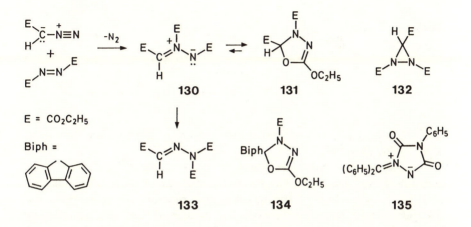

The key role is played by the *azomethine imine* **130**, a zwitterionic species that is not isolable in this case. In 1966 Fahr et al.[144] diagnosed a mobile equilibrium of **134** with an open-chain azomethine imine. The unstable product, obtained from diphenyldiazomethane and 4-phenyl-1,2,4-triazoline-3,5-dione,[145] was recognized in 1965 by Bettinetti and Grünanger[146] as the azomethine imine **135**.

Even earlier (1960) Rudolf Fleischmann, Albrecht Eckell, and I reported the formation of 93% of **136** from 9-diazofluorene and 4-chloro-benzene-*anti*-diazocyanide.[147] The bright orange color and the dipole moment of 6.7 D ruled out a diaziridine structure;[148] an X-ray analysis confirmed the structure **136** (I instead of Cl) of this first azomethine imine.[149] The nitrile stretching frequency at 2118 cm^{-1} corresponds to that of a cyanamide anion. The intermediate on the way to **136** could be either the zwitterion **137** or the cycloadduct **138**.

Stubborn pursuit of a goal is often praised as a virtue, and sometimes leads to success. However, accidental observations can disclose new horizons, far off the original target and sometimes more valuable. The lucky chance might lurk just outside the experimenter's door, but this door is not always opened. Opening it brings *serendipity*— acceptance of Fortuna's gift. The new betaine **136** stood on the shelf for some time before the general scheme of 1,3-dipolar cycloaddition emerged in the Munich laboratory. When the bond system of an azomethine imine was spotted in **136**, it indeed obeyed and performed as an active 1,3-dipole,[150] much to Albrecht Eckell's delight. Rich experi-

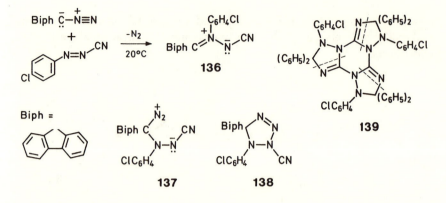

mental support was at hand before we published the cycloaddition concept in 1960 (page 91).

Diazoalkane and diazocyanide were varied;[148] "Eckell's azomethine imine" **136** was easily accessible and stored well. The analogue with $(C_6H_5)_2C$ instead of the spirofluorene residue was converted to a colorless trimer in hot solvents. We presumed that the azomethine imine group cycloadds to the nitrile function of the next molecule and so on; the broken lines in formula **139** suggest how three molecules are pieced together.[151] Later a single-crystal analysis by Isabella Karle substantiated **139**.[152]

Cycloadditions of Azomethine Imines and Their Regiochemistry

3,4-Dihydroisoquinolinium N-arylimides (**141**) were the second track on which azomethine imine chemistry proceeded. In 1958, Ernst Schmitz,[153] well known as the discoverer of diaziridines and diazirines, ascribed a deep red color observed in the hot solution of **143**, Ar = 2,4-dinitrophenyl, to the betaine **141**. We recognized the bond system of our 1,3-dipole in **141**, and Rudolf Grashey prepared more than 100 crystalline adducts of **141** in less than 2 years.[154]

Salts **140** emerge from intramolecular alkylation of aldehyde hydrazones; the red imines **141** are liberated by triethylamine or pyridine and accept dipolarophiles in situ at 20 °C. The dimers **142** exhibit reversible thermochromism above 60 °C and offer a neutral source for **141**; ΔG of dimerization amounts to −6.1 kcal mol^{-1} (R = C_6H_5 in chlorobenzene, 80 °C) as measured by Reinhard Schiffer.[155] The methanol adducts **143** are likewise masked azomethine imines with excellent shelf life.[156]

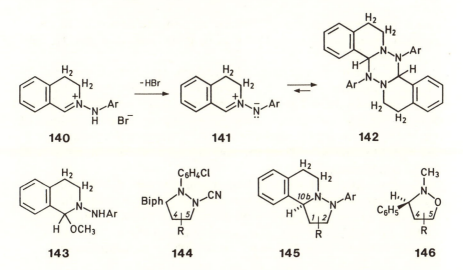

Both **136** and **141** combined with the whole scale of olefinic and acetylenic substrates: enamines, vinyl ethers, common and conjugated alkenes, α,β-unsaturated carbonyl compounds, and nitriles. Rates are higher for **141**, Ar = C_6H_5, than for **136** and reveal more selectivity. Relative rate constants based on 1.0 for 1-hexene reveal for both **136** (at 80 °C) and **141**, R = C_6H_5 (at 50 °C), the minimum function characteristic of nucleophilic–electrophilic 1,3-dipoles (page 106). The k_{rel} of **136** (and **141** in parentheses) are as follows: butyl vinyl ether, 11.6 (156); styrene, 2.8 (27); ethyl acrylate, 55 (30,000); dimethyl fumarate, 12 (106,000); and methyl propiolate, 240 (34,000).[155,157] Andrew Kende and M. V. George participated as postdoctoral fellows in this kinetic study in 1962.

In azomethine imines, the terminal nitrogen is anticipated to hold a larger fraction of the negative charge than the carbon terminus and to be more nucleophilic. Indeed, the reactions of **136** with donor-substituted ethylenes (pyrrolidino, alkoxy, phenyl, and alkyl) provided 5-substituted adducts **144**. The directing force for acceptor substituents to emerge in the 4-position of **144** should be even stronger; a 94:6 ratio of 4- and 5-carboxylic ester in the system **136** + methyl acrylate disclosed a difference of activation free energies ($\Delta\Delta G^{\ddagger}$) of 1.9 kcal mol^{-1}, smaller than expected for the intermediacy of a zwitterion. The 51:49 mixture of regioisomers **145** resulting from **136** and methyl methacrylate, inconsistent with the occurrence of biradical or zwitterion, was another warning shot.[158] We were confronted rather early with the conundrum of *regioselectivity*.[159]

In the cycloadducts **145** from "Grashey's azomethine imine" **141** and ethylenic dipolarophiles, all substituents capable of stabilizing positive charge in the transition state appeared in the 2-position, whereas cyano, methoxycarbonyl, and triphenylphosphonio turned up in the 1-position.[156] Styrene was the switching point and gave rise to both regioisomers of **145**.

The point of regiochemical switching is 1,3-dipole-specific. *N*-Methyl-*C*-phenylnitrone, an azomethine *N*-oxide, combines with olefins to give isoxazolidines **146**.[160] Here not only amino, alkoxy, alkyl, and phenyl, but also ester and nitrile groups appear in the 5-position of **146**.[161,162] As shown by Sims and Houk,[163] only nitroethylene adds in the opposite direction. Thus, the regiochemistry shifts between acrylo-nitrile and nitroethylene.

The addition of **136** to dimethyl fumarate and maleate proceeded with *retention of configuration*. No mutual admixture of cycloadducts was observable in the NMR spectra of the mother liquors.[158] The addition of **141**, Ar = C_6H_5, to fumaric and maleic ester furnished pyrazolidines **147** and **148** in 97% and 95% yield, again without contamination.[156]

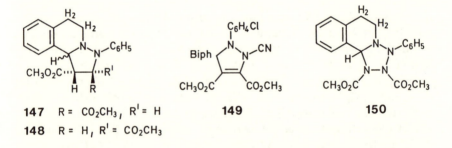

147   R = CO_2CH_3, R' = H   **149**   **150**
148   R = H, R' = CO_2CH_3

The rate constant for the formation of **149** from **136** and dimethyl acetylenedicarboxylate was 3 and 6 times higher in benzene than in acetonitrile or DMF.[157] The small negative response of k_2 to solvent polarity indicates a moderate decrease of charge separation in the activation process. This is anticipated from the measured dipole moments of reactants and **149** if the process is continuous (i.e., if bond formation is *concerted*).

The superior activity of **141**, as compared with **136**, became obvious in cycloadditions to heteromultiple bonds. The C=N bond of phenyl isocyanate and isothiocyanate and the C≡N of ethyl cyanoformate still accepted **136**.[164] Moreover, **141** combined also with carbodiimides, azines, oximes, thiones, and carbon disulfide.[156] Combination with azodicarboxylic ester gave rise to the rare tetrazolidine system **150**. The additions of **141** to aldehydes and aldimines are fast, but nearly thermoneutral; thermochromism points to cycloreversion.[165,166]

In our studies on azomethine imines, X-ray analysis provided insight more than once. The efficiency of this method, achieved through automation and computerization, had tremendous repercussions on organic chemistry. The structural elucidation of low-molecular-weight crystalline compounds by classical methods is no longer a rewarding research goal. My personal attitude is ambiguous. Solving a structural problem by chemical means is an intellectual challenge; X-ray analysis deprived the chemist's activity of such highlights. The rationalist welcomes the saving of time whereas the romanticist deplores the loss of a fascinating game. Playfulness is an incentive for the scientist and a driving force of progress. Be that as it may, I am deeply grateful to X-ray analysts (I. Karle, H. Nöth, A. Gieren, and K. Polborn) for their generous help, especially in difficult configurational clarifications.

X-ray analyses serve various purposes. In the case of **136**,[149] the comparison of bond lengths and angles with models provided information about the nature of the 1,3-dipole. The solubility of trimer **139** was insufficient for ^{13}C NMR spectroscopy, and chemical evidence was unconvincing; in this case, X-ray analysis confirmed our guess.[152] Two more examples from the chemistry of a related azomethine imine may illustrate the invaluable role of X-ray analysis.

An Azomethine Imine with Improper Conduct

The deep-red aromatic isoquinolinium *N*-phenylimide (**151**) is not isolable, but smoothly combines in situ with dimethyl fumarate. Adduct **152** was converted by acid catalysis to an isomer containing a *sec*-amino function. Mechanistic considerations and careful evaluation of the ^{1}H NMR spectrum led Tony Durst, a postdoctoral guest from the University of Western Ontario, Canada, in 1965 to assume structure **154**.[167] The key step is a hydrazo rearrangement involving a [3,3] sigmatropic

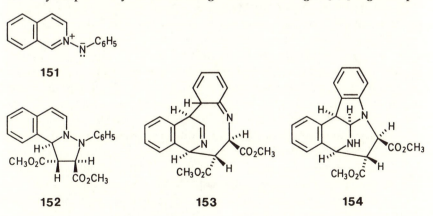

shift **152** → **153**; rearomatization and aminal formation afforded **154**. Robert Temme (Ph.D. thesis 1980), another excellent associate, continued the project and found that many tricyclic adducts of the same type **152** undergo this skeletal rearrangement, which is related to Fischer's indole synthesis. The generality with respect to the substituents in the pyrazolidine ring justified a confirmation of **154** by X-ray.[168]

The adduct of dimethyl maleate, **155**, constituted a bewildering exception to the reaction course discussed. Tony Durst observed in 1965 that **155**, $C_{21}H_{20}N_2O_4$, was converted into yellow crystals of $C_{24}H_{22}N_2O_6$ by picric acid in methanol, the three extra carbon atoms coming from a second molecule of **155**.[167] Twelve years later R. Temme resumed work on the mysterious conversion and assigned all ^1H and ^{13}C NMR signals of the C_{24} compound, but failed to solve the puzzle. On Christmas 1979 an X-ray analysis by Karle and Flippen-Anderson[168] revealed the unanticipated structure **156**.

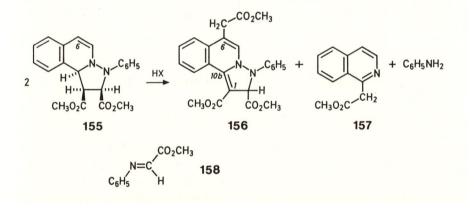

Usually, the X-ray elucidation of a product allows a quick reconstruction of its formation, but not so here. The isolated further products, **157** and aniline, suggested the stoichiometry shown. The decisive step is the acid-catalyzed fragmentation of **155** into methyl isoquinoline-1-acetate (**157**) and glyoxylic ester anil (**158**). The electrophilic **158** attacks the 6-position of a second molecule of **155**; the elimination of aniline is part of an intramolecular redox process leading to the 1–10b double bond of **156**. The adventurous polystep sequence was secured by interception of **158** and model experiments carried out by Jürgen Finke (Ph.D. 1984).[167,169] Miraculously, the reaction of **155** with HCl in dichloromethane at 25 °C was finished after 5 min and yielded 60% of **156**. It was the X-ray analysis that put us on the right track.

"Surprising observation", "not anticipated reaction course"; are there more descriptive terms for our failure to predict the behavior of polyfunctional molecules? Too much of our clarification of reaction pathways still remains *post festum* rationalization; we have good reason for modesty. Not less significant: patent rights are granted only where new procedures have not been predicted. We should not forget, however, that the appeal of organic chemistry springs much from this surprise element.

We envisioned a general access to azomethine imines in the interaction of aldehydes with *N,N′*-disubstituted hydrazines. 4-Chlorobenzaldehyde and *N,N′*-dimethylhydrazine furnished the hexahydrotetrazine **160** as a 2:2 product, whereas combination in the presence of ethyl acrylate produced **162** (12%). The role of the azomethine imine **159** is not unequivocal because polystep sequences shunning **159** are conceivable. Less ambiguous was the reaction of dimer **160** with ethyl acrylate, giving rise to 78% of the pyrazolidine **162**;[170] the thermal dissociation of the related hexahydrotetrazine **142** has been firmly established. Analogous cycloadditions of **160** via **159** with other dipolarophiles were carried out, and carbon disulfide furnished 96% of the 1,3,4-thiazolidine-5-thione **163**.[171]

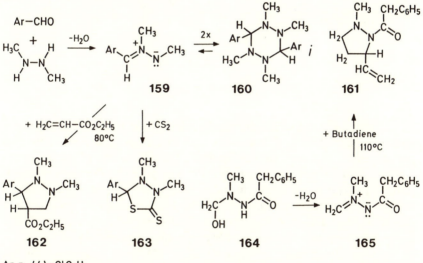

Ar = (4)-ClC$_6$H$_4$

In 1970 W. Oppolzer[172] reacted *N*-methyl-*N′*-phenacetylhydrazine with paraformaldehyde and various types of alkenes, affording pyrazolidines as 1:1:1 products. The azomethine imine **165** was a plausible intermediate, because the hydrazinocarbinol **164** was dehydrated in refluxing toluene. Subsequent combination with alkenes provided cycloadducts (e.g., 68% of **161** with butadiene). Oppolzer[173] elegantly applied this reaction type to intramolecular cycloadditions.

Azodicarboxylic Ester Back Again

The historical role of azo ester **105** as a dienophile has been mentioned, so its quality as a dipolarophile will not be astounding. The additions of 1,3-dipoles at low temperature allowed the preparation of labile ring systems; for example, tetrazolidine **150** harbors four contiguous nitrogens in a saturated ring.

The colorless 1,2,3,5-oxatriazoline **166** precipitated from a solution of benzonitrile oxide and dimethyl azodicarboxylate in ether at −15 °C. At 20 °C, **166** isomerized to the yellow azooxime-O-carboxylic ester **168**. Methoxycarbonyl transfer in the azonium oximate intermediate **167** constitutes a rationalization of the reaction elucidated by Heinz Blaschke, an Austrian postdoctoral fellow. The azo group in **168** accepted a second molecule of benzonitrile oxide, furnishing the symmetrical azobis(O-methoxycarbonyloxime) **169**.[174]

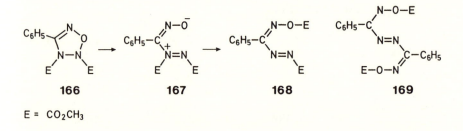

E = CO₂CH₃

The synthesis of 1,3-dipoles should not be restricted to first-row elements. C=S and C=P bonds are well known; why should sulfur and phosphorus not occur in 1,3-dipoles? The chemistry of thiocarbonyl betaines will be outlined on page 110. The adduct **170** from triphenylphosphine and azodicarboxylic ester, mentioned in passing in 1958,[175] formally resembles a 1,3-dipole. In 1969 we designated **170** a quasi-1,3-dipole because **170a** does not contain an electron sextet (*see* page 95) but a phosphonium ion. The labile **170** precipitates in nonpolar solvents.

Several authors preferred structure **171** with attachment of the phosphine to oxygen.[176,177] We favored **170** on the basis of IR data,[178] and in the 1980s ³¹P and ¹³C NMR spectra supported **170**.[179,180] In 1963 Cookson and Locke[177] used formula **171** to interpret the reaction with dimethyl acetylenedicarboxylate (DMAD) affording pyrazole **175** and triphenylphosphine oxide. However, structure **170** likewise explains the formation of **175** via **172–174**. Conspicuously, the carbanion in **172** attacks the ester group and not the phosphonium ion; the elimination of the phosphine oxide from the bicyclic **174** resembles the behavior of 1,2-oxaphosphetanes in the Wittig reaction.

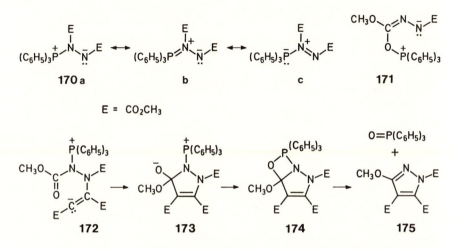

E = CO₂CH₃

On adding azo ester to the solution of triphenylphosphine and phenyl isocyanate or phenyl isothiocyanate, Erwin Brunn (Ph.D. 1968) obtained the triazole derivatives **176**, X = O or S.[181] The analogy with the formation of **175** is obvious.

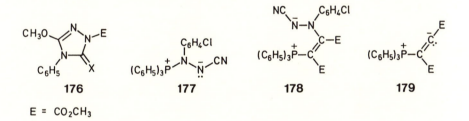

E = CO₂CH₃

Structure **177** was assigned to the pale-yellow crystals from tri- phenylphosphine and 4-chlorobenzenediazocyanide. The reaction of **177** with DMAD gave rise to an orange-red 1:1 adduct. To our surprise, X-ray analysis disclosed formula **178**.[182] Possibly, **177** dissociates and tri- phenylphosphine combines with DMAD to furnish zwitterion **179**, which adds to the electrophilic diazocyanide. Johnson and Tebby[183] established **179** as a short-lived intermediate in the reaction of tri- phenylphosphine with DMAD. This chemistry still belongs to the "unpredictable" category.

In none of the pathways observed so far have the phosphonium betaines **170** and **177** entered 1,3-cycloadditions. In 1971 some reactions of a methyleneaminophosphane and a phosphinyl isocyanate with electron-deficient dipolarophiles were interpreted as (3+2) cycloaddi- tions.[184,185]

In the past 15 years, adduct **170** of triphenylphosphine and azodi-carboxylic ester rose to importance as Mitsunobu reagent.[186] For instance, it affects the formation of carboxylic esters from acid and a chiral alcohol with inversion;[187] triphenylphosphine oxide and hydrazo-dicarboxylic ester result from a *redox condensation*, according to Mukaiya-ma's terminology.[188]

Benzyne Chemistry

Historical Note and Development

In 1874 it was observed that the three isomeric bromobenzenesulfonates were mainly converted to resorcinol by molten alkali hydroxide.[189] Numerous rearrangements in nucleophilic aromatic substitutions that did not fit the accepted addition–elimination mechanism[190] became known and were collected by Bunnett and Zahler[191] in a masterly review (1951). By that time, the clarification of the largest group was on its way.

In 1953 J. D. Roberts et al.[192] reported a brilliant study on the conversion of $[1\text{-}^{14}C]$chlorobenzene to $[1\text{-}^{14}C]$ and $[2\text{-}^{14}C]$aniline by potassium amide in liquid ammonia. The provocative formula **180** was proposed for the intermediate *benzyne*. One year later we published our results[193a] on rearrangements in the aryl fluoride + phenyllithium system, which likewise required an elimination–addition sequence. Our symbol **181** emphasized the undisturbed benzene resonance.

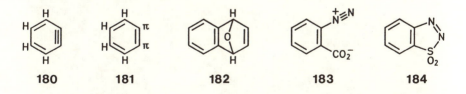

I remember having lunch with a colleague from the Technical University. Clearly he must have thought me crazy, because he asked me whether I expected arynes in my soup. The inclusion of the linear acetylenic bond system into an aromatic ring frightened conservative chemists, but the credibility gap was soon overcome. The endeavors of the Pasadena[193b] and Munich laboratories centered on mechanistic aspects; before long, many more groups joined in the exploration of the new domain.

In nucleophilic substitutions, the aryne usually adds the reagent by which it is generated. In 1956 Wittig and Pohmer[194] uncoupled that sequence of steps by preparing benzyne from 2-bromofluorobenzene via the Grignard compound; benzyne was intercepted by furan as the Diels—Alder adduct **182**, and the one-step synthesis of triptycene from anthracene and **180** followed in the same year.[195]

In 1960 Stiles and Miller[196,197] discovered an access to benzyne that avoided metal organic precursors: the elimination of CO_2 and N_2 from benzenediazonium-o-carboxylate (**183**). Friedman and Logullo[198] contributed a safe variant of in situ preparation of **183**. The fragmentation of **184** offered another, less efficient, source.[199] In 1969 Rees et al.[200] contributed an ingenious new route to benzyne in the oxidation of 1-aminobenzotriazole by lead tetraacetate.

Now all barriers in the choice of reaction partners were removed. Benzyne entered the set of standard reagents, and the broad stream of publications has not dried up since. An excellent book, *Dehydrobenzene and Cycloalkynes* by R. W. Hoffmann (1967),[201] collected the information at an early stage.

How did I get involved? Georg Wittig was a full professor in Tübingen when I joined the staff as an associate professor in 1949. Wittig had used the high basicity of phenyllithium since the 1930s to

With Georg Wittig (1897–1987) (left) at Heidelberg in 1958. Phenyllithium was his tool leading him from one important discovery to the next.

develop an "anionochemistry". The famous Wittig reaction became a late fruit of his systematic approach when phosphonium salts were treated with phenyllithium (1953). Wittig was a great experimenter; careful observation and an uncanny instinct warranted success without his becoming deeply involved with mechanisms. In 1950 he complained about the scant echo of his anionochemistry and encouraged me to join the effort.

Some Rearrangements in Nucleophilic Substitutions

The metalation of benzene derivatives ortho to an acidifying substituent by RLi was independently developed by Gilman[202] and by Wittig in the late 1930s. In 1940 Wittig et al.[203] observed that the *o*-lithiation of fluorobenzene was followed by an amazing substitution, **185** → **186**. Wittig postulated the "dipolar phenylene" **187**, and formula **180** occurred in the same context.[204] However, according to Wittig and Fuhrmann,[205] the analogous reactions of the 12 haloanisoles with phenyllithium proceeded *without rearrangement*. We noticed the discrepancy and, on the spot, Herbert Rist and I observed the positional shifts expected for an elimination–addition sequence.[193]

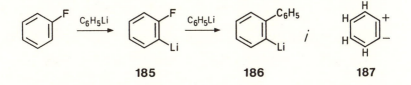

Treatment of 2-fluoroanisole (**188**) and subsequent carboxylation provided 58% of **191** and 4% of **192**. Analytical problems were tough, because no spectroscopic methods were available in German laboratories in the early 1950s. The separation was based on the difference of esterification rate, high-vacuum distillation of the esters, and fractional crystallization. 3-Fluoroanisole (**189**) afforded 80% of **191** and 3% of **192**; these yields suggested 3-methoxybenzyne (**190**) as a common intermediate.[193] Soon afterward IR spectroscopy was at our disposal, and it allowed more precise quantitative analyses.

The reaction of 1-fluoronaphthalene with phenyllithium was likewise accompanied by rearrangement. The intermediate 1-naphthyne accepted phenyllithium in both directions and, after carboxylation, furnished a 63:37 mixture of **193** and **194**. The initial metalation of 2-

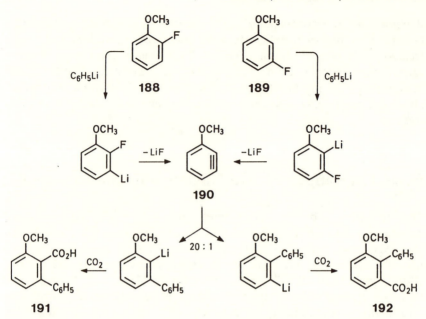

fluoronaphthalene occurred in the 1- and 3- positions (~80:20); 3-phenyl-2-naphthoic acid (195) originated from 2-naphthyne.[193]

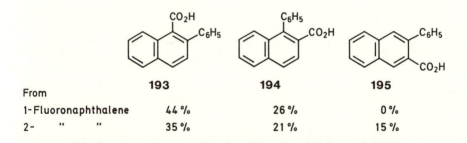

From	193	194	195
1-Fluoronaphthalene	44 %	26 %	0 %
2- " "	35 %	21 %	15 %

One year later Jenny and Roberts[206] reported on the formation of biphenyl from [1-^{14}C]fluorobenzene and phenyllithium after hydrolysis; degradation disclosed 53% rearrangement. The industrial phenol synthesis from chlorobenzene and 4N NaOH at 350 °C takes place mainly via benzyne, as demonstrated with [1-^{14}C]chlorobenzene by Bottini and Roberts.[207]

We converted sodium 2-halophenoxide (Cl, Br, or I) by molten sodium hydroxide at 350 °C into resorcinol + catechol in the constant ratio of 58:42. This conversion suggested sodium benzyne-3-olate (196) as a halogen-free intermediate. Resorcinol and catechol were formed 59:41 from 3-halophenoxides (F, Cl, or Br) under the same conditions, again via the common **196**. Despite the anionic charge on oxygen, the initiating deprotonation of 3-halophenoxide takes place at the 2-position; chelation with NaOH appears feasible. All four sodium 4-halophenoxides gave rise to resorcinol + hydroquinone 75:25, quantitatively proceeding via sodium benzyne-4-olate. Dietmar Jung (Ph.D. 1962) did a superb job.[208]

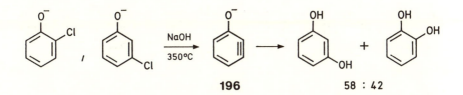

196   58 : 42

Starting in 1945, Gilman studied substitutions of aryl halides by lithium dialkylamides, which were prepared from dialkylamine and 1 equiv of butyl- or phenyllithium in ether; rearrangements frequently occurred. Yields were modest, as exemplified by 18% *N*-phenyl-piperidine (**198**), described by Wittig and Pohmer.[194] We obtained 85–94% of **198** from all four halobenzenes by slowly adding 1 equiv of phenyllithium to the solution of halobenzene and 4 equiv of piperidine.[209] The intermediate **197** competes with lithium piperidide for benzyne and forms products of higher molecular mass; an excess of piperidine quickly protolyzes **197**. By the same procedure, 1- and 2-chloronaphthalene were converted into 32:68 mixtures of 1- and 2-naphthyl-piperidines; only 1-naphthyne occurred in the reaction of 2-halonaphthalene with lithium piperidide.[210] These experiments were part of the Ph.D. thesis of Jürgen Sauer (1957); this competent young associate soon joined me in consulting with the "aryne group".

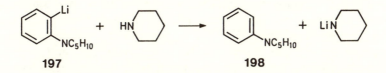

197   **198**

A variety of *N*-heterocycles was prepared from 3-chlorophenyl-alkylamines (e.g., 88% of **201** from **199**, R = H, with phenyllithium via **199**, R = Li, and **200**) by Horst König (Ph.D. 1957).[211] The complete positional change indicates that aryne formation and the selective conversion of **200** are intramolecular.[212]

The preparation of 54% of the 16-membered ring **202** from 12-(3-chlorophenyl)-dodecyl-*N*-methylamine with $NaNH_2$ in benzene testified

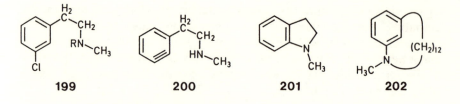

| 199 | 200 | 201 | 202 |

to the usefulness of intramolecular additions to arynes.[213] Hrutfiord and Bunnett[214] initiated the use of potassium amide in liquid ammonia as condensing reagent and effected elegant ring closures.

The Mechanism of Aryne Formation from Aryl Halides

Bergstrom et al.[215] uncovered the reactivity sequence $Br > I > Cl \geq F$ by competition amination of halobenzenes with potassium amide in liquid ammonia. J. D. Roberts et al.[216] established fast D–H exchange of 2-deuteriofluorobenzene in this medium, but the carbanion involved did not lose fluoride ion. For 2-deuteriobromobenzene, $k_H/k_D = 5.5$ suggested a concerted HBr elimination; 2.7 for 2-deuteriochlorobenzene was interpreted as a two-step sequence of deprotonation and chloride loss.[217]

The systems KNH_2–NH_3 and lithium base in diethyl ether—the latter studied in Munich—differ mainly in the inferior aptitude of ether for ion solvation; probably only covalent species and no free anions are involved.

In kinetic measurements by Jürgen Sauer, the first order in aryl halide and lithium base was approximately fulfilled for low conversions (30–40%); the aggregation of lithium bases in ether is a complicating factor. Lithium piperidide reacts 20–90 times faster than phenyllithium with halobenzenes (Table IV). The higher rate of the weaker base—the formation of lithium piperidide from phenyllithium and piperidine proceeds quantitatively—is not consistent with a simple acid–base reaction. Conceivably, the energy of a four-membered cyclic transition state of metalation, 203, is lowered by the presence of an unshared electron pair at the nitrogen of $LiNC_5H_{10}$.

In contrast to the rate sequence observed for KNH_2–NH_3, fluorobenzene ranks at the top (Table IV). This ranking suggests that the metalation step is rate-controlling; k_a decreases from C_6H_5F to C_6H_5I, whereas k_b/k_{-a} increases (Chart IA). Addition of free piperidine to the lithium piperidide diminished the overall k_2 value of C_6H_5F, whereas k_2

Table IV. Rate Constants of Benzyne Formation from Halobenzenes by Lithium Bases in Diethyl Ether

Li Base (equiv)	C_6H_5F	C_6H_5Cl	C_6H_5Br	C_6H_5I
2 LiC_6H_5	4.1	0.40	0.49	0.28
2 $LiNC_5H_{10}$	86	28	45	17
2 $LiNC_5H_{10}$ + 1 HNC_5H_{10}	48	26	73	28
2 $LiNC_5H_{10}$ + 2 HNC_5H_{10}	22	23	115	39
2 $LiNC_5H_{10}$ + 3 HNC_5H_{10}	16	22	150	52

NOTE: All values are rate constants ($10^5 k_2$) in $M^{-1}s^{-1}$. The halobenzenes were at a concentration of ~ 1.8 M and 20 °C. Results were obtained by potentiometric titration of halide ion.

of C_6H_5Cl was not much altered and those of C_6H_5Br and C_6H_5I were increased.[218] The tendency to pass over to the Bergstrom sequence (KNH_2-NH_3) is noticeable.

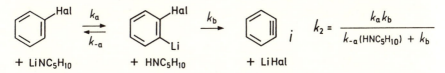

+ $LiNC_5H_{10}$ + HNC_5H_{10} + $LiHal$

Chart IA. Kinetic scheme of benzyne generation from halobenzenes and lithium piperidide.

Hans Meerwein once told me facetiously that it requires intuition to stop an investigation at the right stage. The result of one further experiment can deliver the deathblow to a beautiful hypothesis. The tendency to generalize, to simplify, and to jump to conclusions is inherent to the human mind. Frequently, an investigation has to pass a stage of maximum complexity before reaching the silver-lining rewards for sustained effort.

The results of many more kinetic experiments may be condensed into an enumeration of controlling factors:

1. Piperidine diminishes k_2 by reversing the metalation step; this factor loses weight as k_b/k_{-a} increases from C_6H_5F to C_6H_5I (Chart IA).

2. Table IV offers apparent values of k_2 because the eliminated lithium halide (LiHal) retards the reaction. Therefore, we applied salt-free phenyllithium in order to generate lithium piperidide and measured initial rates with bromo- and iodobenzenes by the

tangent method; the values were ~5 times higher than those listed in Table IV. Careful kinetic evaluation by Wilhelm Mack disclosed ether-soluble 1:1 complexes of LiHal with lithium piperidide, which were inactive in aryne formation; structure 204 was proposed.[219]

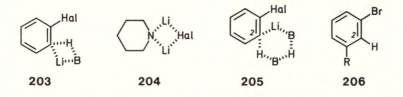

203   204   205   206

3. An excess of piperidine breaks up complex 204 by solvating LiHal. In addition, a six-membered cyclic transition state of metalation, 205, may be more favorable than 203.[218]

4. In 1962 Wittig and Hoffmann[220] showed that halide anions in ethanol add to benzyne, providing halobenzenes. A reversal of the LiHal elimination in ether would belong to the rate-retarding factors.

Does the readiness of aryl halides to undergo o-lithiation correspond to acidities? Experimental k_2 values for aryne formation from substituted bromobenzenes (2 equiv each of $LiNC_5H_{10}$ and HNC_5H_{10} in ether at 20 °C) were dissected into partial rate factors (k_p) for the H vicinal to Br. The substituent influence, measured by Klaus Herbig (Ph.D. 1960), was high for the 2-H of 3-substituted bromobenzenes 206 (relative values based on 0.5 for the 2-H of bromobenzene):[221]

206, R	CH_3	C_6H_5	$N(CH_3)_2$	OCH_3	Br	F
k_p	0.35	1.8	7.3	600	940	1700

The electron-attracting inductive effect of $N(CH_3)_2$ and OCH_3 outplays the retarding electron release to the π system. The exclusive 3-metalation of 2-methoxynaphthalene observed by Sunthankar and Gilman[222] offers elegant support. A satisfactory linear correlation of k_p with the pK_as of substituted pyridinium ions underlines the importance of acidity. The methoxy and dimethylamino compounds did not fit the line; here coordination of RLi at the basic center plays a role.[221]

Characterization of Arynes as Intermediates

"I do not believe in intermediates that cannot be isolated and bottled." Thirty years ago such ingenuousness was not rare. The educated scientist knows that kinetic evidence for reactive intermediates is infallible and can be demonstrated by an arsenal of methods.[223] One criterion relies on a discrepancy between rate and product composition.

Phenyllithium and lithium piperidide compete in the formation of the aryne and in the addition to it. Fluoro- and chlorobenzene were treated with 10–12 equiv of the two lithium bases in varying proportions. The ratio of biphenyl to *N*-phenylpiperidine after hydrolysis was a linear function of the ratio of the two lithium bases. The slope, that is, the competition constant κ, reveals that phenyllithium is 4.4 times faster than lithium piperidide in the product-determining step. In the rate of conversion of fluoro- or chlorobenzene, however, lithium piperidide exceeds phenyllithium 21- or 70-fold (Table IV). Thus, rates and products must be controlled by *different reaction steps*.[224] This rate–product discrepancy was utilized in the piperidine-catalyzed formation of biaryls Ar–Ar' from ArCl + Ar'Li.[225]

Evidence on the nature of the intermediate may come from comparing several precursors. The same value, $\kappa = 4.4$, resulted from experiments with fluoro- and chlorobenzene. Thus, a *halogen-free* intermediate is probably responsible for product formation. As long as precursors are metal-organic, complexes of arynes with metal cations or metal halides cannot be ruled out. Stiles's reagent **183** and **184** are free of that suspicion. Rudolf Knorr (Ph.D. 1963) allowed furan and 1,3-cyclohexadiene to compete for benzyne of different provenance, and the cycloadducts were analyzed by vapor-phase chromatography. Competition constants of 21 ± 2 (Chart 1B) establish the identity of the intermediate and allow only "naked" benzyne.[226]

Arynes were compared in their *selectivity* toward phenyllithium and lithium piperidide. Competition experiments with 1-fluoronaphthalene or 9-chlorophenanthrene unveiled the increasing selectivity of benzyne, 1-naphthyne, and 9-phenanthryne.[224] Is this a function of the decreasing triple-bond distance? The κ values of substituted benzynes (Chart II) intimate that inductive electron withdrawal decreases selectivity, whereas electron release by methyl increases it.[227] The rather demanding measurements of Chart II were carried out by Wilhelm Mack and Leander Möbius. Later Montgomery and Applegate,[228] using the same competition system, added the κ values of cycloalkynes that increase with diminishing deformation of the triple bond: cyclopentyne, 2.6; cyclohexyne, 5.2; and cycloheptyne, 21.

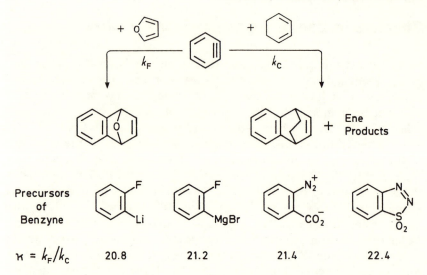

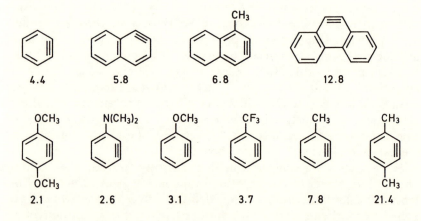

Chart IB. Benzyne as a common intermediate from different precursors; competition reactions with furan and 1,3-cyclohexadiene at 50 °C in tetrahydrofuran.

Chart II. Competition constants κ of arynes for the additions of phenyl-lithium and lithium piperidide in ether.

The lack of a general nucleophilicity scale is a headache for theoreticians, but an inexhaustible source of diversity for the experimenter. *Relative addition rates* of RLi to 9-phenanthryne, based on $LiNC_5H_{10} = 100$, were determined by competition experiments:[229]

R	C_6H_5S	C_6H_5	$N(CH_3)C_6H_5$	$N(C_2H_5)_2$	$O\text{-}tert\text{-}Bu$
k_{rel}	1700	1300	38	26	<3

Lithium thiophenoxide outnumbers phenyllithium in spite of a basicity difference of ~30 pK_a units in the opposite direction; C_6H_5SLi is not even basic enough to liberate 9-phenanthryne from 9-chlorophenanthrene. Virtually identical k_{rel} values of lithium piperidide and free piperidine offered a vexing problem;[230] the basicity difference amounts to ~25 pK_a units.

The two addition directions to nonsymmetric arynes constitute an *intramolecular* competition system; knowledge of the regiochemistry is prerequisite to preparative application. The addition ratio of RLi to 1-naphthyne (**207**), ~65:35 for **208** and **209**, was amazingly constant for R = C_6H_5, *n*-Bu, *tert*-Bu, NC_5H_{10}, $N(C_2H_5)_2$, and N(*iso*-Bu)$_2$. Only very bulky R caused deviations: 87:31 for N(isopropyl)$_2$ and 93:7 for N(cyclohexyl)$_2$.[231] These data were developed by Ludwig Zirngibl (Ph.D. 1957). Addition ratios of nucleophiles to 4-methylbenzyne, collected by R. W. Hoffmann et al.[232] in 1965, cover the small range for *p:m* from 37:63 to 47:53; the data include KNH$_2$–NH$_3$ at –33 °C and aqueous sodium hydroxide at +340 °C.

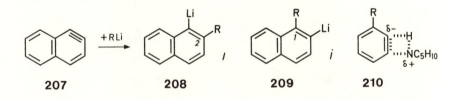

207	**208**	**209**	**210**

The orientations in the addition of piperidine to 3- and 4-substituted benzynes come from experiments, carried out by Klaus Herbig, with 2- and 4-substituted bromobenzenes (Chart III). As expected, 3-substituents command a greater influence than those in the 4-position. The capability of the substituent to stabilize $\delta-$ in the transition state **210** by inductive effect appears to be a directing force besides the steric ortho effect.[233]

I had several reasons for abandoning benzyne chemistry at the beginning of the 1960s. Many groups were active in the field, and other areas like 1,3-dipolar cycloadditions began to blossom in Munich. In addition, Georg Wittig, my venerated senior colleague, signaled in print and word that he regarded dehydrobenzene as his domain.

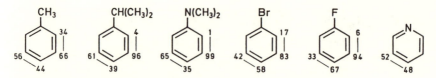

Chart III. Regiochemistry of the addition of piperidine to 3- and 4-substituted benzynes (product ratios with NC_5H_{10} at the marked position); dropwise addition of phenyllithium to aryl bromide and excess of piperidine in ether (infrared analysis).

Cooperating with gifted and enthusiastic students is a privilege. Jürgen Sauer was such a student. He joined the benzyne effort early, was a capable discussion partner, and co-authored two reviews on nucleophilic aromatic substitutions. Christoph Rüchardt and Jürgen Sauer were closely involved in the reform of our laboratory courses in the 1960s. Particularly close ties of friendship developed with my early co-workers.

In his habilitation thesis (Munich 1963) Jürgen Sauer gathered a comprehensive set of kinetic data on Diels–Alder reactions. He greatly contributed to knowledge of the Diels–Alder type with inverse electron demand. Hard work lay ahead for Sauer in 1967: The Institute of Organic Chemistry was to be built at the new Bavarian University of Regensburg. Cycloadditions, electrocyclic reactions, molecular rearrangements, and amphiphilic compounds constitute Sauer's present research interests. (Photograph by Siegfried Hünig.)

Benzyne chemistry may well be the greatest new province of aromatic chemistry to develop in the postwar period. Some facets of its later development may be briefly outlined. Carbocyclic arynes were followed by "hetarynes" derived from six-membered[234,235] or five-membered aromatic compounds.[236] Kauffmann et al.[234] found the selectivity of pyridyne, quinolyne, and isoquinolyne diminished in comparison with the carbocyclic analogs.

Benzyne-type intermediates were heaven for theoreticians. Early Extended Hückel calculations by Roald Hoffmann et al.[237] included daring species like 1,4- and 1,3-benzyne, **211** and **212**. The stability of **212** increases on leaving the frame of the hexagon; there is good experimental evidence[238] for the occurrence of bicyclo[3.1.0]hexatriene (**213**).

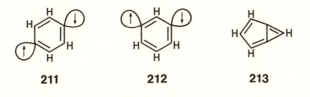

211   **212**   **213**

Azoles and Azides

The Pentazole Story

One hundred years ago it was noticed[239] with amazement that the stability of azoles toward oxidants and acids grew with the number of N atoms (e.g., the benzotriazole system survived refluxing in aqua regia). The "homocyclic" pentazole—all atoms alike—was a fascinating goal, but the synthetic endeavors by Hantzsch,[240] Dimroth et al.,[241] Curtius et al.,[242] and others failed; for example, benzenediazonium chloride and sodium azide furnished phenyl azide + N_2.[240] Each generation tackles certain classical problems anew, after which the problems may fall back into oblivion for a while. When I told Otto Bayer, research director of Bayer AG, in 1955 of our evidence for pentazoles, he noted that sometimes awareness of a problem is half the solution.

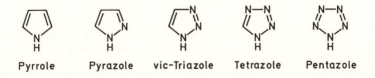

Pyrrole   Pyrazole   vic-Triazole   Tetrazole   Pentazole

I spoke with Ivar Ugi, who had just received his Ph.D. in Munich (1954), about the problem. He observed that the N_2 evolution from benzenediazonium chloride and lithium azide in methanol proceeded in two stages: The "primary nitrogen" (76%) was set free at −40 °C and, after the reaction solution was warmed to 0 °C, the "secondary nitrogen" (24%) followed. Both stages were first-order. The reactants combine in a fast initial step to give benzenediazoazide (214), which, in turn, enters two competing first-order reactions: elimination of the terminal N_2 gives phenyl azide, and ring closure affords phenylpentazole (215). Chart IV separates the upper −40 °C region from the lower 0 °C part. Phenylpentazole, stable at −40 °C, likewise yields phenyl azide + N_2 at 0 °C.

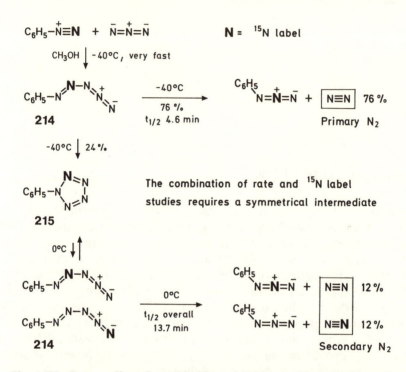

Chart IV. *Benzenediazonium chloride and lithium azide in methanol; N_2 evolution in two stages at −40 °C and 0 °C, and fate of a ^{15}N label.*

This interpretation, published in 1956,[243] found strong support in a ^{15}N-label study in cooperation with Klaus Clusius at Zürich. The primary N_2 did not contain the label, which was introduced as shown in Chart IV, whereas the ^{15}N was diluted fourfold in the secondary N_2.[244,245] When crystalline phenylpentazole was finally obtained at −35 °C, not much additional evidence was gained aside from the UV spectrum, which revealed a modest bathochromic shift compared with 1-phenyltetrazole.[246]

The chemistry of 215 was rather monotonous: N_2 extrusion (half-life = 13.7 min at 0 °C in methanol). Being aware of the equilibrium of imidazides with tetrazoles,[247] Ivar Ugi and I postulated an electrocyclic equilibrium of phenylpentazole with the open-chain **214**; in fact, the term "electrocyclic" was introduced 10 years later.[351] The ring opening of labeled **215** gives rise to two isotopomers of **214**; the half-life of 13.7 min refers to equilibration, **215** $\rightleftarrows$ **214**, *and* N_2 elimination.

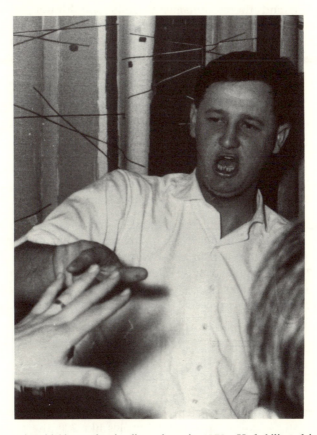

Ivar Ugi, a formidable speaker in discussions, in 1959. He habilitated in Munich (1959), spent six years as research director at the Bayer AG, Leverkusen, was a professor at the University of Southern California (1968–1971), and returned to Munich as a professor at the Technical University. Ugi's unconventional manners and youthful lifestyle won him many friends; his achievements as a dyed-in-the-wool organic chemist are widely respected.

Electron-attracting substituents diminished the half-life of phenylpentazole (4-nitro, 2.0 min in methanol at 0 °C) and electron-releasing substituents increased it up to 69 min for 4-dimethylamino; ρ (Hammett) = 1.0 was based on 13 rate constants. The share of the pathway through arylpentazole ranged from 14% (4-nitro-) to 46% (4-dimethyl-

amino-).[248]   4-Dimethylaminophenylpentazole crystallized from the
methanolic solution at 0 °C. In 1983 Wallis and Dunitz[249] confirmed its
structure by X-ray analysis.

It would be deceptive to ascribe the thermal instability of aryl-
pentazoles to low aromatic resonance energy. Reactivity, a kinetic
phenomenon, and thermodynamic stability should not be confused.
The loss of aromatic resonance in the pentazole ring opening, **215** →
214, is partially set off by the conjugation energy of **214**, and the ensuing cleavage of **214** into phenyl azide + N_2 is exothermic.

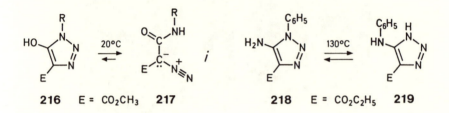

216   E = CO_2CH_3 **217**   **218**   E = $CO_2C_2H_5$ **219**

The chemistry of tetrazoles and 1,2,3-triazoles presents examples
of analogous ring splittings[250] that are inaccessible to benzene-type
compounds. The tautomerism of 5-hydroxytriazoles **216** with diazoacet-
amides **217** was recognized by O. Dimroth in 1910.[251] The isomerization
of 5-amino-1,2,3-triazoles, **218** ⇌ **219**, prototype of the "Dimroth rearrangement" in azoles, proceeds through analogous ring-opened structures.[252]

Ring Opening of 2-Acyltetrazoles

The electron pair at the tervalent nitrogen of tetrazoles is a constituent
of the aromatic π cloud. In N-acyltetrazoles like **220**, the aromaticity is
weakened by a competing carboxamide resonance. N-Acyltetrazoles are
not only active acylating reagents; their propensity for ring opening
should be enhanced.

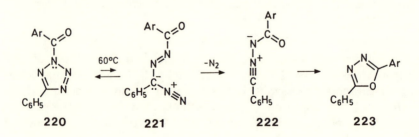

220   **221**   **222**   **223**

The interaction of 5-phenyltetrazole with aroyl chlorides in pyridine at 60 °C afforded the 1,3,4-oxadiazoles **223** in virtually quantitative yields.[253] The life of a research director is not cushioned in velvet. The first experiment with benzoyl chloride in the hands of an inept postdoctoral fellow wound up in black tar. An inner voice told me to have the reaction repeated. Jürgen Sauer, who had just finished his Ph.D., observed complete formation of **223**, even without discoloration of the solution. He could not resist preparing three dozen examples in a few weeks.

The intermediate *N*-acylnitrilimine **222** is subject to 1,5-electrocyclization of the pentadienyl anion type (page 148).[254] The N_2 extrusion from 5-phenyltetrazole and benzoyl chloride in pyridine obeys the first-order kinetics; electron-attracting substituents in the 5-phenyl group decrease the rate and those in the benzoyl group accelerate it.[255] The rate-determining step is either the loss of N_2 from the ring-opened species **221** or a one-step cycloreversion, **220** → **222** + N_2.

A controlled synthesis of oligomers with alternating phenylene and oxadiazole units was built upon the acylation of bistetrazole **224**, available from terephthalonitrile and lithium azide. Interaction with 2 equiv of 4-cyanobenzoyl chloride afforded **225**, *n* = 2. *One* repetition of the two steps put a stop to Hans Jürgen Sturm's ambition; the nonacylic **225**, *n* = 4, mp (dec) ca. 480 °C, was insoluble.[256] Solubility in the *m*-phenylene series was higher, and allowed Christina Axen, a Ph.D. student from Sweden, to continue the iterative procedure up to the 21-nuclear **226**, mp ~380 °C.[257] In the late 1960s the patent literature reflected some interest in these heat-resistant macromolecules, but mainly in the polyreaction, uncontrolled in chain length.

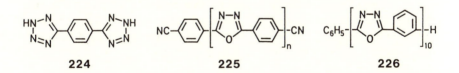

224   **225**   **226**

The weakening of tetrazole resonance by *N*-substituents is a theme in many variations. Treatment with imidoyl chlorides gave access to 1,2,4-triazoles; for example, 98% of **227** from 5-anisyltetrazole and *N*-phenylbenzimidoyl chloride.[258] Michael Seidel (Ph.D. 1960) exploited this reaction. The aromatic 4-chloroquinazoline behaved like an imidoyl chloride toward 5-phenyltetrazole and afforded 96% of **228**; the tetracyclic **229** resulted from cyanuric chloride and three molecules of 5-phenyltetrazole.[259]

Whereas the 2-benzoyltetrazole **220** required 60 °C for the N_2 elimination, 250 °C was needed for the corresponding 1-benzoyl-1,2,3-

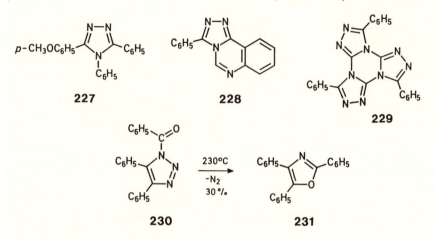

227   **228**   **229**

230   **231**

triazole **230** to furnish the oxazole **231**. The intermediate, corresponding
to **222**, is not octet-stabilized here.[260]

Thermolysis of 2,5-Diphenyltetrazole:
Diphenylnitrilimine

The N atom in the 3-position of phenylpentazole is formally replaced by
$C–C_6H_5$ in the title substance **232**. The N_2 elimination at 160 °C is possibly a one-step reaction, a 1,3-dipolar cycloreversion. The product **233**
represents a new class of compounds and is described by zwitterionic
octet structures. In 1959 we chose the name *nitrilimines* to underline the
relation to nitrile oxides.[261]

When **232** was decomposed in the presence of bases HX, the
elusive diphenylnitrilimine (**233**) was captured and converted to benzoic
acid derivatives (e.g., 73% of the amidrazone **235** was obtained with aniline and 94% of **236** with 4-thiocresol at 170 °C).[262]

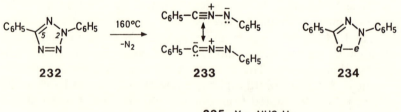

232   **233**   **234**

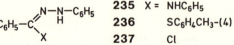

235 X = NHC_6H_5
236 $SC_6H_4CH_3$-(4)
237   Cl

The nitrilimines joined other, newly discovered 1,3-dipoles (page 93). The capability of diphenylnitrilimine (233) to combine with double-bond systems $d=e$ to give five-membered cycloadducts 234 exceeded our expectation. An access to 233 at low temperature appeared desirable. The photolysis of 232 at 20 °C was suitable. Even more convenient was the dehydrochlorination of benzphenylhydrazide chloride (237); slow addition of 1 equiv of triethylamine to the solution of 237 in the presence of a reaction partner generated 233 in low stationary concentration, optimal conditions for interception.[263] Undeniably, 237 is a skin irritant and allergen; sensitive persons—I remember Léon Ghosez as one of the victims—had to move to another room.

Diphenylnitrilimine is not isolable, but competition experiments with pairs of acceptors confirmed the identity of the intermediate 233 involved in all three methods of preparation.[264] The early 1960s were a bonanza. The research on nitrilimines was supported by excellent co-workers who developed the mechanistic aspects and the range of application in an amazingly short time: James Clovis, Albrecht Eckell, Werner Fliege (1941–1990), Jürgen Sauer, Michael Seidel, Reiner Sustmann, and Günther Wallbillich.

A 2,4,6-trimethoxyphenyl on the nitrilimine carbon and a 2,4,6-trinitrophenyl on nitrogen did not have the stabilizing effect we had hoped for; the model underwent a fascinating isomerization.[265] In 1980 the UV spectrum of 233 generated by photolysis of 232 in the matrix at −190 °C was reported by two groups.[266,267] When 233 was produced from 232 by flash pyrolysis at 420 °C, it quantitatively rearranged to 3-phenylindazole, according to Wentrup et al.[268] In 1988 (i.e., 30 years after the discovery of diphenylnitrilimine) Guy Bertrand et al.[269] at Toulouse prepared 238, the first stable nitrilimine (mp 100 °C); admittedly, it was a very exotic one.

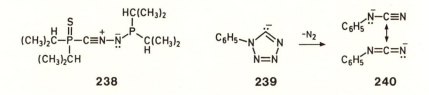

Considerations on the N_2 elimination from pentazoles at low temperature paved the way to nitrilimines. Before we published the general scheme of 1,3-dipolar cycloaddition in 1960,[270,271] diphenylnitrilimine served as a touchstone. We wrote more papers on the chemistry of nitrilimines than on any other class of 1,3-dipoles.

Competition of pairs of dipolarophiles for 233 allowed the elaboration of relative rate constants on the basis of quantitative infrared

analysis, a back-breaking piece of work. Donor *and* acceptor substituents in the ethylenic dipolarophile increased the cycloaddition rate.[272]
Cycloadducts **234** were obtained with a wide range of multiple-bond
systems as dipolarophiles: C=C, C≡C, C=N, C≡N, C=O, C=S, C=P,
N=S, and N=P. The regiochemistry of the additions to CC double and
triple bonds was dictated mainly by the nitrilimine frame, less by the
substituents; the orientational behavior of *C,N*-diphenyl-, *N*-phenyl-,
C-methyl-*N*-phenyl-, and *N*-methyl-*C*-phenylnitrilimine was closely related.[273]

The synthetic potential of in situ cycloadditions of nitrilimines
was widely recognized and exploited. In 1984, a rich literature was bril-
liantly reviewed by Caramella and Grünanger.[274]

A last variation of the pentazole theme: Stollé et al.[275] refluxed
1-phenyltetrazole with aqueous sodium hydroxide and isolated *N*-
phenylcyanamide after acidification. The anion **239** involved is isoelectronic with phenylpentazole (**215**); it loses N_2, and the anion **240** shares
the bond system with phenyl azide.

Literature Browsing; Dibenzamil and the Consequences

The computer age is changing our attitude toward the chemical literature. Search orders with detailed limits are given, and the computer
coughs up citations in abundance. Photocopies of publications—many
of them never read—pile up on the chemist's desk. Many budding
scientists are looking in their research for a niche that offers lifelong
shelter. Overspecialization appears to be nurtured by a flood of new
specialized journals as well as by the library computer service. Who still
has the leisure for browsing? Yet, the old literature is a gold mine for
rediscoveries.

Thermolysis of phenyl azide in inert solvents produces dis-
couraging amounts of tar; aniline and azobenzene were isolated in modest quantities. L. Wolff[276] stated in 1912 that phenyl azide and 1 equiv
of aniline at 150 °C afforded 6–10% of a base $C_{12}H_{12}N_2$. He proposed
the name "dibenzamil" and the tentative structure **241**.

In 1955 the hydrogenation product of dibenzamil was recognized
by Dorothea Vossius as the anil **242** of ε-caprolactam.[277] Active-
hydrogen analysis, as well as IR and UV spectra, suggested **243** for
dibenzamil; the yield was increased to 54% when phenyl azide was
introduced dropwise into 200 equiv of aniline at 165 °C.[278] Innocently
we recommended the name "tropazole" for the new seven-membered
ring system in 1955, until we realized that the *Ring Index* kept the name
"azepine" in reserve.

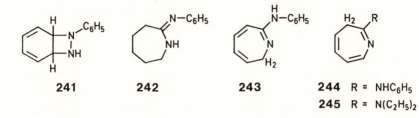

241   **242**   **243**   **244** R = NHC₆H₅

                                                  245 R = N(C₂H₅)₂

Bill Doering has always impressed me as a creative thinker and a superb experimenter. When he visited Munich in 1955, I learned that he likewise knew the structure of dibenzamil, *alias* 2-anilino-7H-azepine. In his 1966 paper[279] he favored the 3H formula **244** on the basis of NMR arguments. He had obtained 34% of **245** on photolysis of phenyl azide in diethylamine.

In the 1950s, labeling with radiocarbon enriched the arsenal of mechanistic tools. We thermolyzed [1-¹⁴C]phenyl azide (**246**) in aniline, and the systematic degradation of **251** by Max Appl located the label in the 2-position.[280]

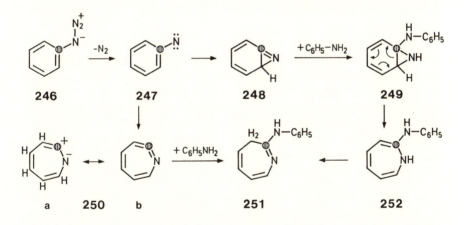

246   **247**   **248**   **249**

    a   **250**   b   **251**   **252**

Did the interaction with aniline occur before or after the ring enlargement? In 1955 we voted for the second possibility with the conversion **247** → **250a**,[277] but 3 years later we reconsidered.[278] Amine addition to the bicyclic 2H-azirine **248** appeared likely, and a "valence tautomerization", **249** → **252**, followed by prototropy should lead to **251**. The one-step process, **246** → **248** + N₂, amounts to an electrophilic aromatic substitution. However, we had to send phenylnitrene (**247**) into the reaction drama because *m*-substituents in phenyl azide hardly influenced the rate of N₂ elimination.[281] Compound **248** looked the more probable as Smolinsky[282] in 1962 reported the preparation of 2Hazirines by thermolysis of vinyl azides. In 1976, quantum chemical calculations supported **248** as a favorable intermediate.[283]

Our investigation was followed in the 1960s and 1970s by a great number of studies on thermolysis and photolysis of aryl azides.[284] In 1968, Crow and Wentrup[285] noticed the interconversion of phenylnitrene and 2-pyridylcarbene on flash pyrolysis (500 °C). In 1978, the pathway via the bicyclic 2*H*-azirine **248** was shown to be incorrect, and our earlier mechanism (1955)[277] was revived. Chapman and Le Roux[286] irradiated phenyl azide in the matrix at 8 K and diagnosed the highly strained ketene imine **250b** by an IR absorption at 1895 cm^{-1}.

In 1955 we lacked the courage to assume **250b** with the linear ketene imine group in a seven-membered ring and preferred the resonance structure **250a** instead. Thirty years later, even [1.1.1]propellanes were not exactly rocking the scientific community. In 1989 we encountered crystalline seven-membered cyclic ketene imines, stable at room temperature (page 119), and in an investigation on (2+2) cycloadditions we see no way past a six-membered ketene imine ring (page 182).

1,3-Cycloadditions of Azides

Although the adduct **253** of phenyl azide to acetylenedicarboxylic ester was described by A. Michael[287] in 1893, cycloadditions of azides received little attention for decades. In 1933 Alder and Stein[288] noticed the high rate of phenyl azide addition to the strained double bond of bicyclo[2.2.1]heptenes. Indeed, norbornene adds phenyl azide at 25 °C 5600 times faster than cyclohexene.[289]

In the early 1960s Raffaello Fusco et al.[290,291] at Milan studied the cycloadditions of aryl and arylsulfonyl azides to enamines; azide sub-

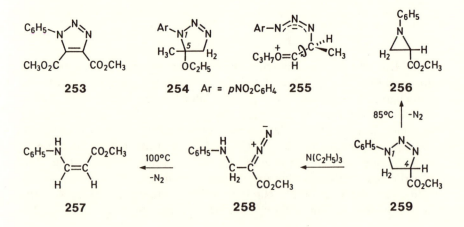

stituent and amino function are located in 1,5-positions of the 1,2,3-triazoline. Additions to vinyl ethers followed the same orientation (e.g., we isolated 99% of **254**;[292] the adverse steric effect is overcome by an electronic directional force. Azides are attacked by nucleophiles at the terminal nitrogen and by electrophiles at the inner one.[293] The >97.5% stereospecificity observed for the cycloadditions of 4-nitrophenyl azide to *trans*- and *cis*-propenyl propyl ethers contraindicated the occurrence of zwitterionic intermediates capable of rotation as **255**.[294] The mechanistic aspects of azide cycloadditions were unveiled in superb work by Leander Möbius and Günter Szeimies.

Phenyl azide also combined with electron-deficient bonds. The reaction with methyl acrylate provided **259** (i.e., the 4-carboxylic ester); the base-catalyzed isomerization to 3-anilino-2-diazopropionic ester (**258**) revealed the regiochemistry.[295] Thus, phenyl azide is bidirectional in

Caution, high strain! All four valences of carbon on one side of a plane in [1.1.1] propellanes and related compounds engineered by Günter Szeimies. This research "propelled" him in 1993 from an associate professorship in Munich to a full professorship at the Humboldt University in Berlin. Szeimies worked as a postdoctoral associate for Ken Wiberg, Yale University, on bicyclobutenes before returning to Munich in 1968. Photo 1980.

addition to donor- and acceptor-substituted ethylenes. The different
pathways of N_2 extrusion are notable. The aziridine formation, 259 →
256, proceeds nonstereospecifically in the case of 4,5-disubstituted 4,5-
dihydro-1,2,3-triazoles.[296] Methyl 3-anilinoacrylate (257) from 258 is
probably formed via a carbene.[295]

Concerted cycloadditions show large negative activation entro-
pies; $\Delta S^{\ddagger}$ values of -26 to -36 eu were measured for phenyl azide additions.[297] Heating is counterproductive because dihydrotriazoles are ther-
molabile; long reaction times at low temperature are recommended. For
example, 4-methoxyphenyl azide and dimethyl fumarate furnished 59%
of the dihydrotriazole after 25 days at 25 °C;[295] admittedly, 13% conversion of cyclohexene + phenyl azide in 13 months is somewhat deter-
ring.[289]

Three N atoms lower the π-MO (molecular orbital) energies of
azides compared with the allyl anion (page 95); the electrophilic character of phenyl azide is more pronounced than its nucleophilicity. Electron-releasing substituents in an ethylenic dipolarophile increase the
cycloaddition rate constant of phenyl azide to a higher extent than
acceptor substituents. Pyrrolidinocyclopentene exceeds cyclopentene
48,000-fold at 25 °C, whereas acrylic ester is only 40 times more active
than 1-heptene. The log k_2 of five aryl azides versus standard dipolarophiles fulfilled Hammett plots, and the ρ values elegantly demonstrated
the dual character of azide reactivity: pyrrolidinocyclohexene, +2.6; norbornene, +0.9; N-phenylmaleimide, −0.8; and maleic anhydride, −1.1.[297]

When the log k_2 of phenyl azide additions are plotted versus the
ionization potentials of the dipolarophiles, a U-shaped curve—
somewhat lopsided toward high ionization potentials—resulted. This
curve served as prototype for a simple, but very successful, perturbation
molecular orbital (PMO) theory model of reactivity sequences in concerted cycloadditions developed by my former co-worker R. Sustmann
in 1972 and discussed on page 106.

1,3-Dipolar Cycloadditions

With 94 full papers, 109 communications, and 28 review articles on various aspects, 1,3-dipolar cycloadditions constitute my most extensive research effort; approximately 80 full papers are still to be written. The mechanistic considerations leading to the general concept in 1958 were recounted at the decennial of the *Fonds der Chemischen Industrie* (Foundations of the Chemical Industry), 1960[270] and in a Centenary Lecture to the Chemical Society (1960).[271] The multiauthor two-volume book *1,3-Dipolar Cycloaddition Chemistry* (1984) provides a first-rate survey. The editor, Albert Padwa, himself very successful in the field, gave me the opportunity of outlining historical development and mechanistic analysis in the introductory chapter.[298]

The *Science Citation Index* reflected wide interest. I found myself among the 250 scientists most quoted in 1961–1976 and still so in the 1980s.[299] "1,3-Dipolar Cycloaddition, Past and Future",[159] published in *Angewandte Chemie* (1963), was my most-cited paper. A second 1963 review on kinetics and mechanism[300] appeared on the list of 100 chemical articles most cited in 1972.[299] "Always the wrong ones," scribbled my friend Manfred Schlosser on the margin beside no. 7 of a list comprising "The 49 most cited organic chemists for the period 1965–1978", when he sent me a copy.[301] Perhaps the timing was significant; it came as a surprise that a rich domain was terra incognita for so long. The volume of the experimental work helped to convince fellow chemists of the synthetic potential of 1,3-dipolar cycloadditions.

Heterocyclic chemistry is a labyrinth of incredible dimensions. Numerous specific syntheses of each heterocyclic system defy simple ordering principles, and their incoherence does not help to win the sympathy of textbook authors and students. The acceptance and appre-

"...venturing experiment from the stepping-stone of mechanistic considerations."
From the laudation of the Liebig Medal awarded by Egon Wiberg (right),
President of the Gesellschaft Deutscher Chemiker *in Aachen in 1961.*

ciation of 1,3-dipolar cycloadditions may have arisen partly because a single synthetic scheme gave access to *many five-membered heterocycles.*

The pioneering discovery of (4+2) cycloadditions by Diels and Alder[121] (1928) was followed by beautiful systematic work of the Alder school at Cologne. The dream of reactions proceeding quantitatively under mild conditions without need of catalysis is often fulfilled by concerted cycloadditions. The retention of reactant configuration allows *stereocontrol* over as many as four terminal centers; this situation is reminiscent of Emil Fischer's paradigm of lock and key fitting. In the past 15 years, the *intramolecular* variant of the Diels—Alder reaction and 1,3-dipolar cycloaddition gained particular importance in the stereo-selective synthesis of natural products.[302,303]

In this chapter, brief historical comments will be followed by less well-reviewed phenomena, and some recent progress will be annotated.

Chemistry and Classification of 1,3-Dipoles

Beyond a prime interest in mechanisms, I was fascinated by the possibility of using mechanistic insight to improve synthetic methods, create new ones, and uncover new classes of compounds.

The development of a new research area proceeds slowly until a phase of exponential growth is reached. In the first year one co-worker was engaged in preliminary exploration, mainly kinetic measurements. In the second year two more joined the effort; then, in the third year, 17 co-workers reaped the harvest of 1,3-dipolar cycloadditions. In this context I would like to mention the role of my able assistant, Rudolf Grashey, as a pacesetter.

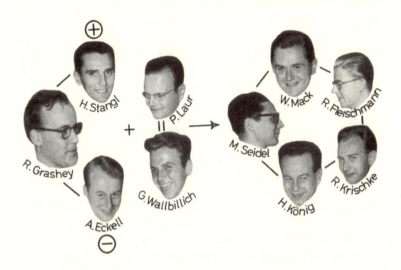

1960 Vintage of players in 1,3-dipolar cycloaddition. (Reproduced with permission from the commemorative publication of Fonds der Chemischen Industrie. *Copyright 1960 FCI, Düsseldorf.)*

The structure–rate relationship for the cycloadditions of various diazoalkanes reflected ground-state stability and not nucleophilic or electrophilic reactivity. The addition rates of phenyl azide to bicyclo[2.2.1]heptene derivatives showed a minute or even negative response to solvent polarity. These observations were reason enough to assume *concerted* formation of the two new σ bonds. Why should this process **260**, with cyclic electron shift, be limited to diazoalkanes and azides? The isoelectronic relationship of R_2C, RN, and O, as well as that of RC and N, allowed us to visualize many new 1,3-dipoles; it was a challenge to make them accessible. The lectures of 1960[270,271] presented four novel classes to illustrate the general principle: nitrile ylides, nitrilimines, azomethine ylides, and azomethine imines. New cycloadditions of nitrile oxides and nitrones served as a supplement.

Thus, a mechanistic consideration led to the general concept of 1,3-dipolar cycloaddition, and mechanistic studies accompanied its development.

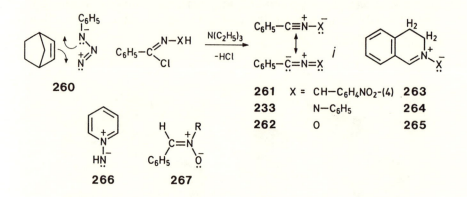

260

261 X = CH—$C_6H_4NO_2$-(4) **263**

233   N—C_6H_5 **264**

262   O   **265**

266   **267**

The first representatives of the new classes of 1,3-dipoles were not the parent compounds, but conveniently accessible derivatives. The synthesis of diphenylnitrilimine (**233**) from benzphenylhydrazide chloride (**237**) and triethylamine (pages 84–85) found its analogy in generating benzonitrile N-p-nitrobenzylide (**261**) from the imidoyl chloride;[304] **261** was not isolable, but smoothly combined in situ with dipolarophiles. Cycloadditions of benzonitrile N-oxide (**262**) were studied by A. Quilico and his school.[305] The dimerization of **262** giving 4,5-diphenylfuroxan is an unwelcome competitor. The liberation of **262** from benzhydroximoyl chloride by slow addition of triethylamine in the presence of dipolarophiles substantially improved the procedure.[306] This simple stratagem of a low stationary concentration of **262** was soon so commonly used that citing our paper with Wilhelm Mack appeared unnecessary. Manfred Christl found this technique mandatory for the 1,3-cycloadditions of fulminic acid set free from formhydroximoyl iodide.[307]

The azomethine imines **136** and **264**, and their cycloadditions[150,154] were mentioned earlier (pages 55–59). Similarly, Elmar Steingruber prepared the first azomethine ylide **263** by deprotonation of 3,4-dihydro-N-p-nitrobenzylisoquinolinium bromide with triethylamine and reacted it in situ with dipolarophiles.[308] Roland Krischke opened up the 1,3-cycloadditions of pyridinium N-imide (**266**) and of analogous derivatives of quinoline and isoquinoline. The overcoming of the aromatic resonance revealed a high driving force.[309] The N-oxide **265** and open-chain nitrones like **267** served in the hands of Rudolf Grashey as azomethine oxides to probe the wide realm of dipolarophiles.[160]

In the search for scope, limitations, and regiochemistry of the new cycloadditions, a multitude of adducts was prepared and structurally elucidated. Their number was estimated to exceed 500 in 1960[271] and 1000 in 1963.[159]

In 1960 the definition of the 1,3-dipole was based on the minor resonance contributor of **268** and **269**: a system *abc* with an electron sextet and concomitant positive charge at *a* and an unshared electron pair at the anionic center *c*. That definition also comprised 1,3-dipoles **270** with "external octet stabilization" and six 1,3-dipoles **271** "without octet stabilization"; the latter possess carbon as middle atom.[271]

In 1968, systems **270** and **271** were eliminated from the list, and the term 1,3-dipole was confined to 4π systems of the propargyl–allenyl type (**268**) and allyl type (**269**).[310]

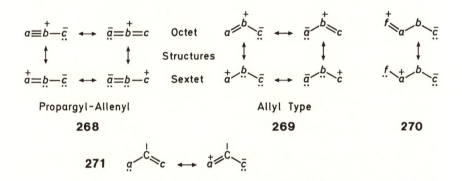

This 4π system of the allyl anion must be the reacting unit; in a 1963 review this was deduced from experimental facts.[300] The description of the *two-plane orientation complex* of 1,3-dipole and dipolarophile, as well as the movements in the activation process, fully accorded with the later Woodward–Hoffmann treatment of concerted cycloadditions (1965)[311] and did not require alteration.

Chart V records the orthodox 1,3-dipoles built from C, N, and O as center atoms and illustrated by one of the octet resonance structures. Nine of the 18 classes were unknown previously, and cycloadditions had been reported with only five of them. Today, nitrosimines and nitrosoxides are still unknown, but 1,3-cycloadditions have been carried out with all the others. Inclusion of elements of the second row of the periodic table, especially phosphorus and sulfur, increases the number of 1,3-dipoles considerably; recent results on S-containing 1,3-dipoles are summarized here. Numerous compounds harboring the bond system of the allyl anion type do not conform to **268** and **269** in the distribution of formal charges; their cycloadditions were reviewed in 1963[159] and 1984.[298h]

1,3-Dipoles of Propargyl-Allenyl Type

Nitrilium Betaines		Diazonium Betaines	
$-C\equiv\overset{+}{N}-\overset{-}{\underset{\cdot}{C}}<$	Nitrile ylides	$N\equiv\overset{+}{N}-\overset{-}{\underset{\cdot}{C}}<$	Diazoalkanes
$-C\equiv\overset{+}{N}-\overset{-}{\underset{\cdot}{N}}\diagup$	Nitrile imines	$N\equiv\overset{+}{N}-\overset{-}{\underset{\cdot}{N}}\diagup$	Azides
$-C\equiv\overset{+}{N}-\overset{-}{\underset{\cdot\cdot}{O}}$	Nitrile oxides	$N\equiv\overset{+}{N}-\overset{-}{\underset{\cdot\cdot}{O}}$	Dinitrogen oxide

1,3-Dipoles of Allyl Type

Nitrogen Function as Middle Center		Oxygen Atom as Middle Center	
$>C=\overset{+}{\underset{\vert}{N}}-\overset{-}{\underset{\cdot}{C}}<$	Azomethine ylides	$>C=\overset{+}{O}-\overset{-}{\underset{\cdot}{C}}<$	Carbonyl ylides
$>C=\overset{+}{\underset{\vert}{N}}-\overset{-}{\underset{\cdot}{N}}\diagup$	Azomethine imines	$>C=\overset{+}{O}-\overset{-}{\underset{\cdot}{N}}\diagup$	Carbonyl imines
$>C=\overset{+}{\underset{\vert}{N}}-\overset{-}{\underset{\cdot\cdot}{O}}$	Nitrones	$>C=\overset{+}{O}-\overset{-}{\underset{\cdot\cdot}{O}}$	Carbonyl oxides
$\diagup N=\overset{+}{\underset{\vert}{N}}-\overset{-}{\underset{\cdot}{N}}\diagup$	Azimines	$\diagup N=\overset{+}{O}-\overset{-}{\underset{\cdot}{N}}\diagup$	Nitrosimines
$\diagup N=\overset{+}{\underset{\vert}{N}}-\overset{-}{\underset{\cdot\cdot}{O}}$	Azoxy Compounds	$\diagup N=\overset{+}{O}-\overset{-}{\underset{\cdot\cdot}{O}}$	Nitrosoxides
$O=\overset{+}{\underset{\vert}{N}}-\overset{-}{\underset{\cdot\cdot}{O}}$	Nitro Compounds	$O=\overset{+}{O}-\overset{-}{\underset{\cdot\cdot}{O}}$	Ozone

Chart V. Classification of 1,3-dipoles with C, N, and O as center atoms.

In the 1960 concept, the 1,3-dipoles **268** and **269** were illustrated by one octet and one sextet resonance structure.[271] The sextet structure contributes little, but symbolizes the 1,3 reactivity. The term "1,3-dipole" may have had suggestive power; soon I felt annoyed by the widespread use of sextet formulae, for which originally I had to bear the blame. In physical terms the 1,3-dipole complies with the definition of a quadrupole. Houk and Yamaguchi[312] provided a competent theoretical discussion of the bond system. According to our definition[310] (1968), a 1,3-dipole is a compound *abc* that undergoes 1,3-cycloadditions and is described by zwitterionic octet structures.

Many 1,3-dipoles are capable of and giving three-membered rings by electrocyclization. The thermal equilibration of suitably substituted aziridines and oxiranes with azomethine ylides and carbonyl ylides, respectively, will be outlined later (pages 139–148). The photolysis of 2*H*-azirines offered convenient access to nitrile ylides, as discovered by Padwa and Schmid.[313]

Chart V of 1,3-dipoles ends with ozone. Its reaction with alkenes is a standard laboratory procedure. Many features of Criegee's ozonolysis mechanism (1953)[314] became clear in 1963 when we assumed *concertedness* for the three steps: 1,3-dipolar cycloaddition of ozone to the $C=C$ double bond, 1,3-dipolar cycloreversion to carbonyl oxide + carbonyl compound, and renewed 1,3-cycloaddition of the fragments to give 1,2,4-trioxolanes.[159] Criegee formulated the larger fragment as the "zwitterion", a sextet structure, which was recognized only many years later as a carbonyl ylide capable of *syn,anti* isomerism. Correct observations could have avoided error-stricken escapades on the mechanism of ozonation by other groups in the 1960s. Rudolf Criegee (page 196) was a venerable personality; his last paper,[315] published in 1975, reestablished the three-step pathway without fanfare.

Intermezzo: Acylnitrenes and Acylcarbenes

Our 1963 review[159] still recorded 1,3-cycloadditions of sextet species **271**: ketocarbenes, thioketocarbenes, and ketoazenes; the latter are called acylnitrenes today. Although 1,3-additions of these 2π species probably proceed in two steps, our search may be briefly reviewed here.

The N_2 extrusion from diazoketones **272** to give ketocarbenes **273** may be initiated thermally, by silver ion catalysis, or by light. The Wolff rearrangement of **273** to the ketene **274** is usually fast.[316] Our attempts at intercepting benzoylphenylcarbene by dipolarophiles in the thermolysis of azibenzil failed. We devised tricks to slow down the rearrangement: use of substituents R with low migratory aptitude, complexation with metal, and incorporation of the C—C bond of **273** in a benzene ring.

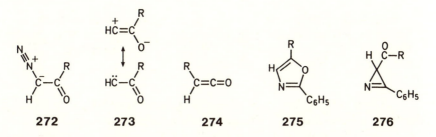

| 272 | 273 | 274 | 275 | 276 |

When diazoacetophenone (**272**, R = C_6H_5) was heated in benzonitrile at 150 °C, only 0.4% of 2,5-diphenyloxazole (**275**, R = C_6H_5) was isolated as interception product. Electron-attracting substituents R retarded the sextet rearrangement, **273** → **274**, and the percentage of **275** rose to 38% for R = 4-chlorophenyl and 45% for R = 4-nitrophenyl in experiments with substituted diazoacetophenones,[317] as observed by

Léon Ghosez on a brief postdoctoral stay in Munich. Copper accelerates N_2 extrusion from **272**, but slows down the Wolff rearrangement. Gerhard Binsch[318] found that, in the presence of copper powder or Cu(II)—acetylacetonate, diazoacetophenone in benzonitrile yielded 16% of **275**, R = C_6H_5.

Gerhard Binsch (1934–1993) in 1963. Originally a technician at BASF, he received the Ph.D. in Munich (1963). After years at Caltech and Notre Dame, Binsch returned to Munich as a full professor in 1972. An autodidactic theoretician, much of his work was in NMR theory.

Wolff rearrangement of ethoxycarbonylcarbene (**273**, R = OC_2H_5) has been observed only in the *photolysis* of ethyl diazoacetate (**272**, R = OC_2H_5).[319] We isolated 42% of **275**, R = OC_2H_5, from the thermolysis (150 °C) of diazoacetic ester in benzonitrile.[320] The nagging doubt remained: Could **275** be formed via the 2*H*-azirine **276**? Thermal isomerization of 2-acyl-2*H*-azirines to oxazoles has indeed been observed.[321] Doering and Mole[322] studied the formation of cyclopropenes from **273**, R = OC_2H_5, and C≡C triple bonds. The cyclopropanation of olefins with diazoacetic ester is a standard procedure, and complexing with copper increases the selectivity in favor of the 1,1-cycloaddition.

On irradiation, 2-diazophenoxides **277** are converted to the ring-contracted ketenes **279** via **278**; the azo coupling of the intact **277** is the basis of an inexpensive technique of photoreproduction.[323] Our expectation that Wolff rearrangement of **278** would be retarded by loss of aromaticity proved correct. Thermolysis and photolysis of 2-diazotetrachlorophenoxide (**280**) were studied by Gerhard Binsch in his Ph.D. thesis (1963). In benzonitrile at 130 °C, **280** was converted to the benzoxazole **281**. Heating of **280** in carbon disulfide or phenyl isothiocyanate at 140 °C yielded the benzoxathiole derivatives **282**, X = S and X = NC_6H_5, respectively.[324] The dipolarophiles did not interfere with the rate-determining N_2 expulsion, as a kinetic study by Horst König revealed.

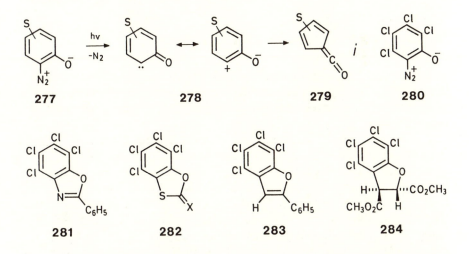

The reaction of **280** with acetylenic and olefinic dipolarophiles allowed the isolation of 1,3-cycloadducts (e.g., with phenylacetylene, **283**, and with dimethyl fumarate, **284**). The same *trans*-diester **284** resulted from the interaction with dimethyl maleate and suggested an intermediate capable of rotation.[325] Consideration of the electronic configuration of **278** raised doubts with respect to the concertedness of these cycloadditions.[324]

Thermolysis or photolysis of ethyl azidoformate produced ethoxycarbonylnitrene, which combined with cyclohexene to give the bicyclic aziridine **285**;[326] the conversion of benzene into azepine-1-carboxylic ester (**286**) was a breakthrough achieved by Hafner and König.[327] On heating of ethyl azidoformate in diphenylacetylene, Heinz Blaschke, a postdoctoral fellow from Graz, obtained 2-ethoxy-4,5-diphenyloxazole (**287**, 33%); the normal rate of N_2 elimination excluded an initial azide cycloaddition.[328] Are we dealing with a one-step 1,3-cycloaddition or a

1,1-addition affording the antiaromatic 1*H*-azirine **288**, which subsequently rearranges to **287**? The generation of ethoxycarbonylnitrene by photolysis in acetonitrile furnished the 1,3,4-oxadiazole **289** (60%).[329]

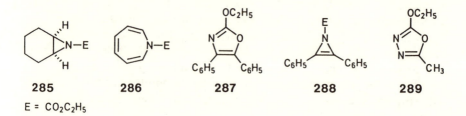

285   **286**   **287**   **288**   **289**

E = CO₂C₂H₅

Mesoionic Heterocycles as 1,3-Dipoles

Sydnones are 1,2,3-oxadiazolium-5-olates that were discovered in Sydney in 1935. David Ollis persistently developed the chemistry of meso-ionic heterocycles, intellectually (definition, classification) as well as experimentally.[330]

The two resonance formulae chosen in **290** for *C*-methyl-*N*-phenylsydnone contain the octet structures of an azomethine imine (Chart V) embedded in a five-membered ring with six π electrons. The first tests were instantly successful, and in 1962 Rudolf Grashey, Hans Gotthardt, and I reported on 1,3-dipolar cycloadditions of sydnones to alkynes and alkenes.[331,332] Kinetic studies established a rate-determining addition step followed by a fast elimination of CO_2, as illustrated in **291** and **292** (82%) for the reaction of **290** with ethyl phenylpropiolate.[333] The interaction of **290** with methyl propiolate producing the pyrazoles **294** and **295** in a 65:35 ratio denotes an *ambident* nucleophile.[334]

In the reaction of sydnone **290** with alkenes (e.g., styrene), a bicyclic intermediate analogous to **291** suffers 1,3-dipolar cycloreversion, affording CO_2 and a new azomethine imine **296**. The latter lacks aromaticity and is stabilized by prototropy or a symmetry-allowed [1,4] hydrogen shift furnishing the 2-pyrazoline **297** (79%). Intermediates of type **296** can also accept a second molecule of dipolarophile giving bisadducts.[335]

The driving force of these cycloadditions astounds. When sydnone **290** was allowed to react with 1,1-diphenylethylene in refluxing xylene, 40% of pyrazole **293** was formed and benzene was eliminated![336]

The wealth of synthetic applications can be illustrated here by only a few examples. Rudolf Grashey, whose pacesetter role was already mentioned, did the initial tests and then graciously left the field

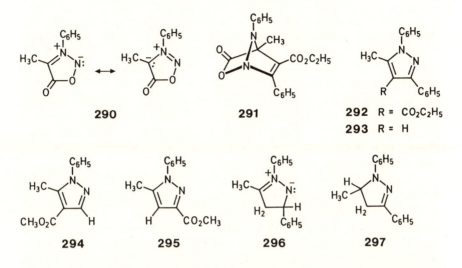

to Hans Gotthardt, who built his dissertation upon the new reactions of sydnones. In his first year as a graduate student (1961), he stunned me with the results of an experiment we had not discussed before. Oxazolin-4-ones (azlactones) like **298** are anhydrides of *N*-acylamino acids and play a key role in the chemistry of amino acids and peptides. On refluxing either **298** or **299** with dimethyl acetylenedicarboxylate in xylene, Gotthardt obtained the pyrazole-3,4-dicarboxylic ester **301** in high yield.[337] The inertness of 4,4-dimethyl-2-phenyloxazolinone (**298**, 4-CH$_3$ instead of 4-H) suggested that a tautomer was the 1,3-dipole. The mesoionic oxazolium-5-olate **302** differs from the oxazolinone **300** by a 1,2 hydrogen shift.

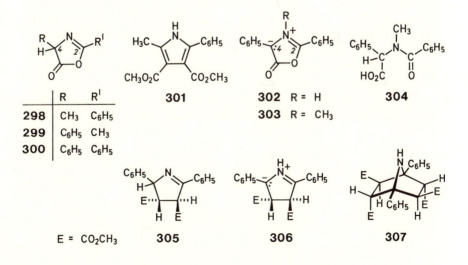

Short treatment of N-benzoyl-N-methylphenylglycin (304) with acetic anhydride yielded yellow crystals of 303.[338,339] With the sydnone nomenclature in mind, my co-workers nicknamed 302 and 303 "münchnones", and the name stuck. The colorless 300 gave yellow solutions; comparison of the extinction with that of 303 suggested a tautomeric equilibrium, 300 ⇄ 302, the fraction of 302 being 50% in DMF, 0.3% in acetone, and 0.01% in chloroform.[340] The feared racemization in peptide synthesis has been attributed to the formation of azlactones; there is a good chance that the tautomeric equilibrium with the münchnone of type 302 is responsible.

A splendid trio of postdoctoral fellows, imaginative and skilled, quickly developed the chemistry of the novel oxazolium-5-olates: Horst O. Bayer, coming from the Rohm and Haas Company of Philadelphia; Hans Gotthardt, who stayed for another year after obtaining his Ph.D.; and Fred C. Schäfer, on a leave of absence from the American Cyanamide Company in Stamford. The richness of the harvest (20 papers 1964–1971; for the last ones some more hands helped) makes it difficult to design a sketch of the colorful münchnone reactivity.

Formula 302 reveals the bond system of an azomethine ylide (Chart V). Reaction of 300 with 1 equiv of dimethyl fumarate at 120 °C furnished the 1-pyrroline 305 (67%), whereas introduction of 300 into an excess of molten dimethyl fumarate gave rise to the bisadduct 307 (76%). The tautomer 302 undergoes cycloaddition to fumaric ester, and the bicyclic adduct loses CO_2 in a dipolar cycloreversion. The emerging nonaromatic azomethine ylide 306 tautomerizes to 305; if the concentration of dimethyl fumarate is sufficiently high, 306 is intercepted by the second cycloaddition affording the 7-azabicyclo[2.2.1]heptane 307.[341]

Oxazolinone 300 reacts via a tiny equilibrium concentration of the mesoionic tautomer 302. The interaction of münchnone 303 with methyl propiolate at 0 °C was free of this bottleneck and produced the tetrasubstituted pyrrole 308 (94%) by the addition–elimination sequence. General syntheses of pyrroles and the otherwise not easily accessible 2-pyrrolines were based on the high 1,3-dipolar activity of münchnones versus acetylene and ethylene derivatives.[342,343]

Alkyl substituents in the 2- or 4-position of 303 foil the isolation of the oxazolium-5-olate because of the ensuing Dakin–West reaction.[344]

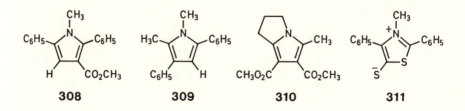

308   309   310   311

However, treatment of an *N*-acyl-*sec*-amino acid with acetic anhydride in the presence of a dipolarophile allows *in situ* cycloadditions of the initially formed münchnones. *N*-Benzoyl-*N*-methylalanine warmed with phenylacetylene and acetic anhydride at 80 °C afforded 87% of the pyrrole **309**. It seems a miracle that 76% of the bicyclic pyrrole **310** emerged from the one-pot reaction of proline, acetic anhydride, and dimethyl acetylenedicarboxylate.[342]

Cycloadditions of oxazolium-5-olates to heteromultiple bonds were manifold. Carbon dioxide bubbled out of a solution of **303** in carbon disulfide, and the orange-red thiazolium-5-thiolate **311** was quantitatively formed.[345]

In the reactions of **303** with aldehydes, thioketones, and electrophilic azo compounds, the sequence of cycloaddition and cycloreversion was followed by another pericyclic step: electrocyclic ring opening of the cyclopentenyl anion type (pages 148–149).[254] Nitrosobenzene was converted by **303** at 20 °C into 97% of the *N*-benzoylamidine **312**; the dashed line in the formula separates the former dipolarophile.[346]

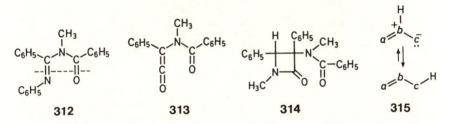

| **312** | **313** | **314** | **315** |

Münchnone **303** itself entertains an electrocyclic equilibrium with the open-chain ketene **313**. The reaction of **303** via **313** with *N*-benzylidenemethylamine led straight to the 3-acylaminoazetidinone skeleton of the penicillins: Eberhard Funke (Ph.D. 1967) obtained 62% of **314**.[347]

I was thrilled by this chapter of cycloaddition chemistry, and I was not the only one. Several groups further exploited the cycloadditions of mesoionic systems, among them my former associate Hans Gotthardt, who became a professor at the new University of Wuppertal in 1976; he studied mainly sulfur-containing mesoionic rings. Gotthardt's death in 1988 prematurely ended a promising scientific career. He was one of the best experimenters I ever met.

K. T. Potts at Rensselaer Polytechnic Institute has likewise been active in the area for 20 years. A review of the field, which he wrote in 1984, listed more than 200 references.[348] The field was vastly expanded by the inclusion of 1,3-cycloadditions of heteropentalenes and a plethora of further mesomeric betaines; Ollis et al.[349] provided a new typology. Of course, as far as $(3+2 \rightarrow 5)$ cycloadditions are involved, the abundance is reduced to the 1,3-dipoles of Chart V.

Reticent but highly efficient in research: Hans Gotthardt (1932–1988) in 1968. Gotthardt had a difficult youth with resettlement from the German Southeast to the German Democratic Republic in 1946 and escape to West Germany in 1950. After five years as technician in chemical industry, Gotthardt studied chemistry at the University of Munich in record time.

A concluding note: the tautomerism **300** ⇄ **302** corresponds to scheme **315** and is a prototype of a growing number of examples.[298g] The recent research by Grigg et al.[350] deserves mention.

The Mechanistic Question

The allyl anion is a strong nucleophile. A good 1,3-dipole requires a proper balance of nucleophilic and electrophilic character within the same molecular framework. The electrophilicity emanates from the onium center *b* and becomes effective in *a* and *c*, as shown in formulae **268** and **269**. It is not meaningful to assign fixed electrophilic and nucleophilic termini to a 1,3-dipole, a truism for symmetrical species like ozone or azomethine ylides. A literature search produced evidence for the *ambivalence* (i.e., potential electrophilic and nucleophilic character at each terminus) of diazoalkanes, azides, and nitrile ylides.[298b]

With the common π electronic description, $_\pi 4_s + _\pi 2_{s'}$ the principle of conservation of orbital symmetry stressed the kinship of Diels–Alder reaction and 1,3-dipolar cycloaddition.[351] All the criteria for *concertedness*—stereospecificity, the small influence of solvent polarity on the rate, *trans–cis* rate ratios ≥ 1, large negative activation entropies, and kinetic isotope effects—are similar for $(4+2)$ and $(3+2)$ cycloadditions and were reviewed.[298,300] Personal experience weighs more than literature reports. Before me lay a legion of specimens in glistening crystals, all these new cycloadducts accessible under mildest conditions. I felt fascinated by the idea that 1,3-dipolar cycloadditions owe their very wide scope to the *concerted pathway* that avoids high-energy intermediates. There is experimental evidence for early transition states (TS), whereas those of a two-step process should resemble the intermediate.[298c]

Hard nuts to crack were the U-shaped *reactivity sequences*; often electron-releasing *and* -withdrawing substituents increase the activity of an ethylenic dipolarophile. It was recognized early that concerted bond formation does not imply "marching in step" (i.e., the same extent of bonding for the two incipient σ bonds in the TS).[300] Inequality should result in partial charges in the TS that are stabilized by substituents. The magnitude of the rate effects, however, defies an explanation by partial charge stabilization.

Furthermore, the regiochemistry created headaches. Some 1,3-dipoles, such as diazoalkanes, azides (page 89), and azomethine imines (pages 56–59) are *bidirectional* (i.e., they add in opposite directions to donor- and acceptor-substituted ethylenes). On the other hand, nitrilimines, nitrile oxides, and nitrones tend to be more *unidirectional*; donor and acceptor substituents appear in the same ring position. A closer scrutiny brought to light the fact that no 1,3-dipole is truly unidirectional; the point of regiochemical switching may shift toward the electron-deficient or electron-rich side of the dipolarophile scale.[298f,352] "The orientation phenomena in 1,3-dipolar as well as Diels–Alder addition offer perhaps the biggest unsolved problem in the field" was my statement of 1968.[310]

The bond energy of C–C, C–N, and C–O exceeds that of N–N, N–O, and O–O, respectively. In 1963 it appeared to me that additions to heteromultiple bonds are guided in a direction that gives rise to new strong C–hetero bonds rather than weak hetero–hetero bonds.[300] This "principle of maximum gain of σ-bond energy" was later riddled, and I discarded it in the light of better insight. I tried not to cling to preconceived ideas, and many of the 1963 interpretations[300] were later abandoned.

The scales fell from my eyes when my former co-worker Reiner Sustmann expounded a simple model in 1971 using perturbation MO

(PMO) theory to interpret the *reactivity sequences*.[353,354] The difference in the energy gains ΔE_I and ΔE_{II} (Figure 2) offers the key to the 1,3-dipole-specific U-shapes of the reactivity sequences; the U-shapes that resulted when log k_2 was plotted versus the electron density of the dipolarophilic π bond. It was an aesthetic delight to see the reactivity profiles gradually change in going from the nucleophilic diazomethane[355] to the increasingly electrophilic diazoacetic ester, diazomalonic ester, and diazo(phenylsulfonyl)acetic ester.[356] These kinetic measurements by Jochen Geittner (Ph.D. 1974) and Hans-Ulrich Reissig were in qualitative accord with the Sustmann concept.

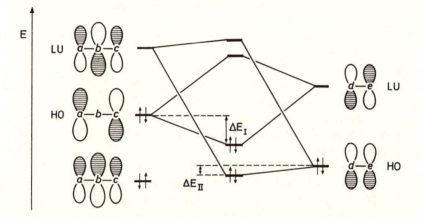

Figure 2. PMO correlation diagram (Frontier π MOs) of 1,3-dipolar cycloaddition. The energy levels in the middle refer to the transition state.

The energy gains $\Delta E_I + \Delta E_{II}$ calculated for frontier orbital (FMO) interactions on the basis of the PMO model should be proportional to the log of the rate constants. A breakdown of the FMO treatment was noted for the additions of diazomethane to conjugated alkenes, which are very active dipolarophiles. According to Sustmann's calculations, a fairly linear correlation with log k_2 was achieved only when HO(diazomethane) was allowed to interact with all unoccupied MOs of the olefinic dipolarophile.[357] One of the reasons is as follows: The compression of the highest occupied–lowest unoccupied (HO–LU) gap in conjugated systems cannot outweigh the diminution of the atomic orbital coefficients in the second-order term of the PMO equation. Conjugated ethylenes harbor *more* MOs than the parent; thus the restriction to FMOs does not hold.

Modern MO calculations are highly successful in reproducing properties of stable molecules. An unsatisfactory phase of theoretical chemistry, in which experimental data were confirmed *post festum* by quantum-mechanical calculations, appears to be overcome. The credibility of the method was not improved when eventual revisions of experimental figures were followed by an adequate revision of calculated values.

Today computational chemistry is revealing its potential in simulating molecular properties and in predicting the structures of unknown molecules. The rise of theoretical chemistry is tied to increasing computer efficiency. The numerical values still should be taken with a grain of salt because changes in the basis set, type of calculation, and applica-

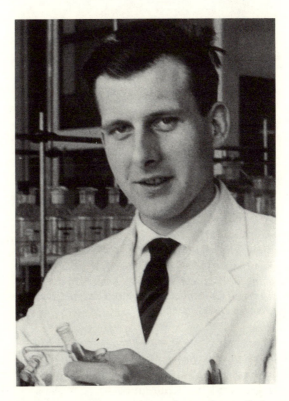

A creative mix of theory (PMO) and experiment (free radicals, cycloadditions, transition metal mediated reactions) are the scientific contributions of Reiner Sustmann. Sustmann, pictured as a Ph.D. student in 1963, became a professor at the new University of Essen in 1978.

tion of configuration interaction substantially influence numerical results. Development of more reliable structures and energies of transition states appears feasible as a central goal for the decade to come.
Michael Dewar, Roald Hoffmann, Ken Houk, John Pople, and Paul von
R. Schleyer all made pioneering contributions in the application of
quantum mechanics to problems of organic chemistry. Molecular mechanics offers a tool of phenomenal value, considering its relatively
modest demands on computer time.

*Paul von R. Schleyer started with computational chemistry in the early 1970s and
became one of its most potent and successful advocates. In 1974 he received the
Senior Research Award of the Humboldt Foundation and stayed at the Munich
Institute, indulging in the use of the Leibniz Rechenzentrum. It was a great
gain for chemistry in Germany when Schleyer moved from Princeton to the
University of Erlangen (1976).*

In 1964–1965 my friend Paul Schleyer spent a sabbatical year in
Munich. He impressed me as a splendid discussion partner; I admired
his intellectual grip and his willingness to cooperate. I often tap his
encyclopedic knowledge and experience in computational chemistry.

Bastide et al.[358] and Houk et al.[359] exploited the potential of PMO for the interpretation of *regioselectivity* of 1,3-dipolar cycloadditions. The problem is tricky because $\Delta\Delta G^{\ddagger} = 1.0$ kcal mol^{-1} is sufficient to generate an 85:15 ratio of regioisomers at 25 °RC. Qualitative predictions are safe when both FMO interactions favor the same orientation. Otherwise the calculation should involve more than frontier MOs. Recently Sustmann[360] emphasized the importance of variations in the repulsive first-order term of the perturbation equation. Among various regiochemical problems, the direction of diazomethane addition to ethyl vinyl ether was calculated correctly for the first time.

A spin-paired *biradical intermediate* was considered and discarded in 1963[300] as a reason for the unidirectional cycloadditions of diphenylnitrilimine (233) to olefins; however, methyl propiolate gave rise to two regioisomers. Suspecting general unidirectionality, Ray A. Firestone[361] in 1968 postulated biradical intermediates for all 1,3-dipolar cycloadditions. "The strong tendency for each 1,3-dipole to add in the same direction to both electron-rich and electron-poor olefins"[362] is a basic misconception.

The assessment of a mechanistic idea highly profits from a divergence of opinions. All the arguments for concertedness are inevitably *indirect* and based on the exclusion of two-step mechanisms. Without an opponent, we may not have searched for an example with >99.997% stereospecificity (page 115). We would not have prepared a Firestone-type biradical. Its behavior was at variance with that postulated by the critic; it did not dissociate and it cyclized nonstereospecifically.[352] The arguments for the occurrence of biradicals in 1,3-dipolar cycloadditions and Diels–Alder reactions were carefully refuted;[310,352] when the old arguments were reiterated unchanged in 1977 by my opponent,[363] constructive correspondence, already filling a file, came to a halt. I met with Ray Firestone several times in the 1980s; during our friendly meetings we discussed many common interests—never, and never again, touching the subject of cycloadditions. The biradical hypothesis did not find many zealots among scientists working in the area, but it confused the chemical community.

Personally, I have never regarded it as a loss of face to change my mind. The scientist searching for the "eternal laws of nature" is an offspring of some journalist's overheated fantasy. Our goal is not *truth*, a philosophical term, but rather a *consistent description and interpretation* of phenomena within a model concept. New observations may call for alteration or even replacements of such models.

A witty colleague proposed the strict separation of the descriptive and interpretative parts of a publication; the interpretation should be printed on perforated paper, ready for removal after some decades. I do not agree; I find it enlightening to watch the change of our views on

reaction mechanisms. Mechanistic theories are not a novelty of our
time, as is often falsely assumed. The founding fathers, Liebig, Wöhler,
Dumas, and Kekulé, were inspired by the wish to understand reactivity.
Physical chemistry, still to be developed, would bridge the gap between
physics and chemistry.

Thiones as Superdipolarophiles

My first contact with organosulfur chemistry concerned the clarification
of the "Schönberg reaction". In 1930 Bergmann et al.[364] and Schönberg
et al.[365] independently discovered the formation of tetraphenyl-1,3-
dithiolane (317) from thiobenzophenone and diazomethane at 0 °C.
Patient variation of the reactants by Schönberg et al.[366] over several
decades did not clarify the mechanism.

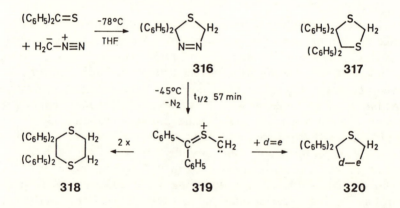

Thiobenzophenone reacted with diazomethane in THF at −78 °C
at a rate at which titration was still feasible, but now the stoichiometry
was 1:1 and the thiadiazoline 316 crystallized. The elimination of N_2 at
−45 °C followed first-order kinetics, and the head–head dimer 318 was
isolated. Thiobenzophenone S-methylide (319), the first representative
of the class of thiocarbonyl ylides occurred as a short-lived intermediate
and was intercepted in situ by a second equivalent of thiobenzophenone (in the role of d=e), affording 95% of 317. Thus, the
Schönberg reaction consists of three steps: 1,3-dipolar cycloaddition of
diazomethane to the C–S double bond, 1,3-dipolar cycloreversion of
316, and renewed cycloaddition of thiocarbonyl ylide 319 to the thione.[367] Ivars Kalwinsch, a guest of the Latvian Academy of Sciences,
solved the problem with bravura. A small digression: My description,

that the crystals of **316** "went pffft" around −20 °C, was not approved of by the referee, but the editor of the *Journal of the American Chemical Society* did not object to a small dose of humor.

When other dipolarophiles were added to the solution of **316** at −78 °C, the reaction at −45 °C furnished an abundance of 1,3-dithiolanes, thiolanes, and dihydrothiophenes in high yields.[368] The competition of pairs of dipolarophiles for **319** at −45 °C and HPLC analysis of the cycloadducts **320** allowed the determination of relative reactivities. Based on many pair connections, Xingya Li[369] established a string of relative rate constants (Li spent 2 years in Munich as a guest of the Shanghai Institute of Organic Chemistry). Chart VI offers a selection of k_{rel}. Dipolarophiles less electrophilic than methyl propiolate did not combine with **319**; that observation points to HO(1,3-dipole) + LU(dipolarophile) as the dominant interaction in the TS.

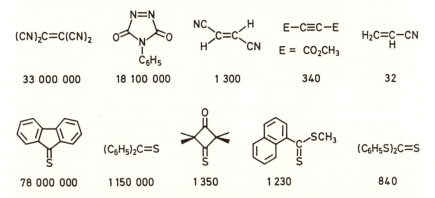

Chart VI. Relative rate constants of thiobenzophenone S-methylide (319), based on methyl propiolate ≡ 1 (THF, −45 °C).

The respectable range of 10^8 for the k_{rel} of the highly nucleophilic thiocarbonyl ylide **319** reveals an unusual selectivity. The stepwise introduction of three CN groups into acrylonitrile is accompanied by a millionfold rate increase that reflects the lowering of the LU energy. Less clear are the high rates of thioketones (Chart VI) because the CS double bond is not electron-deficient. Thiofluorenone exceeds even TCNE, and thiobenzophenone reacts 3000 times faster than dimethyl acetylenedicarboxylate (DMAD).

The titration of thiobenzophenone with a solution of diazomethane at −78 °C furnishing **316** suggests again a fabulously high dipolarophilic activity of thioketones. Elke Langhals[370] chose diphenyldiazomethane, still a nucleophilic 1,3-dipole, for rate measurements (Chart VII). Diazoalkanes are notorious for their propensity to add to α,β-unsaturated esters and nitriles. Once more thiones are super.

Thiofluorenone beats TCNE by a factor of 115 and thiobenzophenone is 1300 times more active than ethyl acrylate.

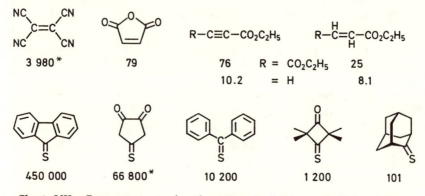

3 980*	79	76	R = CO₂C₂H₅	25
		10.2	= H	8.1

450 000	66 800*	10 200	1 200	101

Chart VII. Rate constants for the 1,3-cycloadditions of diphenyldiazomethane to electrophilic alkenes and thiones; VIS spectrophotometry 10^3 k_2 in DMF (CDCl₃) at 40 °C.*

Nitrones are nucleophilic–electrophilic 1,3-dipoles. In accordance with Sustmann's PMO model,[354] they react fast with electron-deficient dipolarophiles, slow with common alkenes, and again fast with electron-rich double bonds like enamines. In dilatometric rate measurements with N-methyl-C-phenylnitrone (**267**, R = CH₃), Helmut Seidl established a table of dipolarophilic activities with DMAD at the top position. These 1969 kinetic data[371a] were recently supplemented by Lubor Fisera, a guest from Bratislava (Chart VIII).[371b] Even the highly encumbered 2,2,6,6-tetramethylcyclohexanethione ranked 5 times higher than DMAD, and adamantanethione is 1500 times faster, despite a sizable steric hindrance. The influence of solvent polarity on k_2 is small, thus ruling out zwitterionic intermediates.

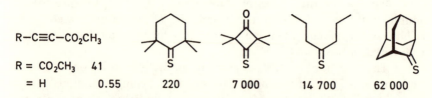

R–C≡C–CO₂CH₃				
R = CO₂CH₃ 41				
= H 0.55	220	7 000	14 700	62 000

Chart VIII. Cycloaddition rates of N-methyl-C-phenylnitrone with acetylenic carboxylic esters and thiones; VIS spectrophotometry, 10^5 k_2 in toluene at 25 °C.

Why are thiones superdipolarophiles? The efficient PMO model appears to offer the key. We ascribe the top-notch activity to a diminished HO–LU energy distance of the CS π bond compared with the CO π bond. This phenomenon is supported by calculations as well as by the wavelengths of the $\pi \rightarrow \pi^*$ transitions. One or both ΔE values of Figure 2 are increased, and the attractive second-order term of the perturbation equation lowers the TS of the cycloaddition. The low HO–LU gap may likewise be responsible for the high polarizability of the CS double bond.

Thiones are also highly active dienophiles in the Diels–Alder reaction. The notoriously unstable thioaldehydes were intercepted by 1,3-dienes. Recently Sauer and Schatz[372a] confirmed the high reactivity of thiofluorenone and thiobenzophenone by rate measurements. The low HO–LU energy distance should be a general phenomenon for π bonds of elements of the second or higher long period. High cycloaddition rates were indeed qualitatively shown for compounds with CSi double bond, CP triple bond, etc.

Thiocarbonyl *S*-oxides (sulfines) were known to be efficient dipolarophiles,[372b] but were failures as 1,3-dipoles. The thione stratagem worked yet again. Sulfine **320a** and thiobenzophenone equilibrated at room temperature with the 1,2,4-oxadithiolane **320b**. As expected, the new ring system was stable in the cycloadduct **320c** obtained from **320a** and adamantanethione.[372c] The X-ray analysis of the dispirobis(adamantane) **320d** revealed a dihedral angle C–S–O–C of 55°; the lone-pair repulsion is responsible for the twist at the S–O bond. It was Grzegorz Mloston, a guest from the University of Lodz, who overcame the resistance of sulfines.

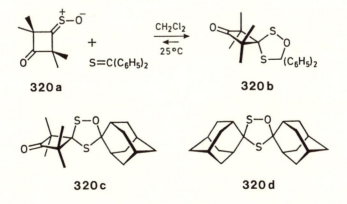

Thiocarbonyl *S*-sulfides are new 1,3-dipoles that have to be reacted in situ. Two moles of thiobenzophenone reacted with 2,2-

diphenylthiirane in pentane (14 days, 25 °C), affording 92% each of 1,1-diphenylethylene and 3,3,5,5-tetraphenyl-1,2,4-trithiolane (**321**). Diphenylthiirane and thiobenzophenone both entered the rate equation with first order; the second mole of thiobenzophenone was consumed in a fast subsequent 1,3-dipolar cycloaddition of thiobenzophenone S-sulfide (**322**).[373]

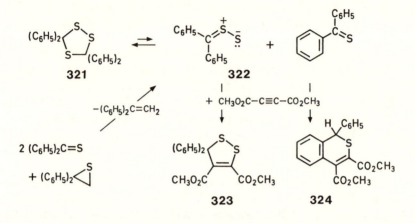

The colorless solutions of **321** turn blue, the color of thiobenzophenone, at 60 °C. When the dissociation equilibrium of the trithiolane **321** was established in the presence of DMAD, the latter acted both as a dipolarophile and as a dienophile, furnishing the yellow $3H$-1,2-dithiole **323** (83%) and the $1H$-2-benzothiopyran **324**.[373] The cycloadditions of **322** to thiones are the fastest, however. In the formation of **321** from **322** + thiobenzophenone the kinetics is fine; we are cornered by thermodynamics (equilibrium). In the reactions of **322** with aliphatic thiones (e.g., adamantanethione), the equilibria are far on the 1,2,4-trithiolane side. The chemistry of thione S-sulfides was developed by Jochen Rapp (Ph.D. 1988), the son of Walter Rapp, one of my first graduate students.

Stereospecificity

Retention of configuration is a necessary, but not conclusive, criterion of concertedness. Two-step processes may appear stereospecific if the rate ratio of rotation versus cyclization of the intermediate biradical or zwitterion is sufficiently small (Chart IX). Retention was observed for dozens of cycloadditions with *cis–trans* isomeric dipolarophiles or 1,3-dipoles[298d] (i.e., the amount of nonstereospecific product remained below the analytical limit). If the NMR spectrum of artificial mixtures of *cis* and *trans* cycloadduct allows us to detect 2% of the minor isomer and the latter is missing in the cycloadduct, then only >98% stereospecificity can be guaranteed, corresponding to $\Delta\Delta G\ddagger = 2.3$ kcal mol^{-1} for the two-step process in Chart IX.

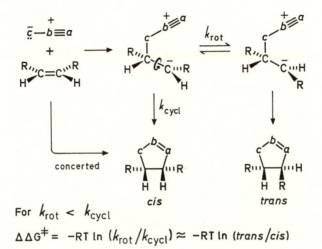

$$\text{For } k_{rot} < k_{cycl}$$

$$\Delta\Delta G^{\ddagger} = -RT \ln (k_{rot}/k_{cycl}) \approx -RT \ln (trans/cis)$$

Chart IX. Stereospecificity in cycloadditions may originate either from concertedness of bond formation or from a small ratio of rotation rate versus cyclization rate of an intermediate.

In search of much higher stereospecificity, we wanted to apply gas chromatography (GC) as an analytical tool. The cycloaddition of diazomethane to methyl tiglate (**325**) quantitatively afforded the *cis*-dimethyl-1-pyrazoline **326**. As little as 30 ppm of the *trans*-dimethyl isomer could be detected by GC in artificial mixtures. The *trans* peak was missing in the cycloadduct, a result demonstrating a retention that exceeded 99.997%.[374] In the framework of a two-step mechanism, this value would impose an improbably high rotational barrier (>6.2 kcal mol^{-1}) on the intermediate. Three of my young associates, one after the other, improved the precision of the capillary GC in the 1970s; the result described here was achieved by Hans-Ulrich Reissig.

In 1979, Dorn et al.[375] claimed the first violation of the retention principle in a 1,3-dipolar cycloaddition; 15–30% inversion product (at C-2) was obtained along with **328** in the reaction of 1-benzylidene-pyrazolid-3-one betaine (**327**), an azomethine imine, with *trans*-β-nitrostyrene. Under identical conditions, k_{rot}/k_{cycl} of an intermediate should be constant. The reported span made me skeptical. After meeting with Dorn, we tried to reproduce "the first non-*cisoid* 1,3-dipolar cycloaddi-

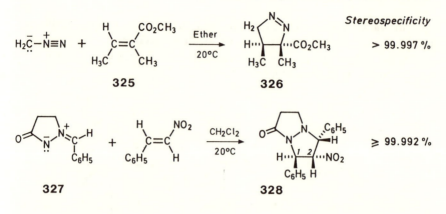

325   **326**   *Stereospecificity* > 99.997 %

327   **328**   ≥ 99.992 %

tion". *sec*-Nitroalkanes are stronger acids than phenol. Rudolf Weinberger (Ph.D. 1989) found that a 66:34 equilibrium of **328** with the 1,2-*cis* isomer was attained (CDCl$_3$, 0 °C) in the presence of traces of base. New cycloaddition experiments under *highly sterile* conditions provided **328** with very high retention; HPLC allowed a precise analysis.[376]

In 1985, Houk, Firestone, et al.[377] described an experiment based on known ratios k_{rot}/k_{cycl} of tetramethylenes **329**. The ratio is 12 for **329**, R = D, and ~1 for R = CH$_3$; no rotation preceded the ring closure of a tetramethylene with two terminal radicals C(CH$_3$)C$_2$H$_5$. Thus, a *prim*-alkyl terminus should likewise offer the highest chance of detecting the

The consequences of the PMO treatment of cycloadditions were examined by Hans-Ulrich Reissig in his Ph.D. thesis (Munich 1978): rate profiles, reactions of electron-deficient diazo compounds with enamines, 3H-pyrazoles. His habilitation thesis (Würzburg 1984) dealt with substituted siloxycyclopropanes as C$_3$ synthon. In 1993, after an interlude at Darmstadt, Reissig accepted a full professorship at Dresden, Saxony. The snapshot shows Reissig and me at a lab party in 1975.

rotation of an eventual biradical intermediate in 1,3-dipolar cycloaddition. However, the formation of isoxazoline **330** from 4-nitrobenzonitrile oxide and *cis*-1,2-dideuterioethylene proceeded with ≥98% retention, 2% being the analytical limit. A reasonable conclusion was that no intermediate capable of rotation occurs on the pathway to **330**.

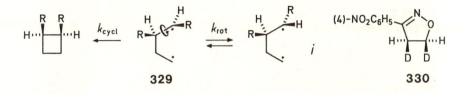

329   **330**

Borderline Crossing: Two-Step Mechanism

A PMO consideration led us to the first two-step 1,3-cycloadditions. In Figure 2 different energy distances of the interacting FMOs are responsible for $\Delta E_I > \Delta E_{II}$ (i.e., the energy gains in the TS). Will ΔE_{II} not become negligibly small if the energy gap between LU(1,3-dipole) + HO(dipolarophile) is further increased? It may be so small that the *additional entropy bill* of the concerted process cannot be settled any more. The now-unidirectional electron flow establishes one bond furnishing a zwitterion; the process should be promoted by steric encumbrance at one of the 1,3-dipole's termini.

We pondered how to provide a 1,3-dipole with high π MO energies. The usual 1,3-dipoles (Chart V) harbor nitrogen or oxygen; these elements of higher electronegativity lower MO energies. Among heteroatoms, sulfur has the same electronegativity on the Pauling scale as carbon (2.5). Formal replacement of the middle CH of the allyl anion by a sulfonium function gives rise to *thiocarbonyl ylides*; thiobenzophenone S-methylide (**319**) is a representative of this class of highly nucleophilic 1,3-dipoles. An ethylene with four electron-attracting substituents would fit the role of a dipolarophile with very low MO energies (electrophile). In such a pair of 1,3-dipole + dipolarophile, ΔE_{II} (Figure 2) might be negligible compared with ΔE_I.

1,3,4-Thiadiazolines derived from aliphatic thiones and diazoalkanes are more stable and easier to handle than those derived from aromatic thiones. When thiocarbonyl ylide **332** was set free from the precursor **331** at 45 °C in the presence of dimethyl 2,3-dicyanofumarate or dimethyl 2,3-dicyanomaleate, kinetically controlled cycloadditions gave rise to mixtures of *trans*- and *cis*-thiolanes, **335** and **336**.[378] Yet be prepared for traps. Thiadiazoline **331** was shown to catalyze a slow

Does the spectrum give the desired information? Grzegorz Mloston and I were wondering in 1991. As an Alexander von Humboldt Fellow from the University of Lodz, Poland, in 1983–1985, Mloston participated with skill and fervor in the studies of thiocarbonyl ylides. On four further stays in Munich, 1988–1993, his output stayed at high levels: 15 papers and many more to come. His habilitation at Lodz dates from 1991, and he has been a professor since 1992.

cis–trans equilibration of the dipolarophiles (*trans,cis* 92:8 at equilibrium). Grzegorz Mloston[379], a postdoctoral associate from the University of Lodz and an excellent experimenter, overcame the obstacle: A trace of concentrated sulfuric acid in $CDCl_3$ (7.6 μmol mL^{-1}) and a higher reaction temperature (85 °C) suppressed the *cis–trans* isomerization of the unsaturated diesters. The thiolanes **335** and **336** are configurationally stable, hence their ratios in Chart X demonstrate a non-stereospecificity effected by rotation of the zwitterionic intermediates, **333** and **334**.

These intermediates **333** and **334** contain *tertiary termini*, and yet they rotate. We can hardly avoid the conclusion: Stereospecific 1,3-dipolar cycloadditions follow a fundamentally different mechanism, a concerted one.

The reaction of thiocarbonyl ylide **332** with tetracyanoethylene in THF + 2 vol% methanol produced the seven-membered lactim ether **337** along with some thiolane. A strong case was built by Grzegorz Mloston for the fleeting occurrence of a highly strained cyclic ketene imine as an intermediate that was captured by methanol or water.[380]

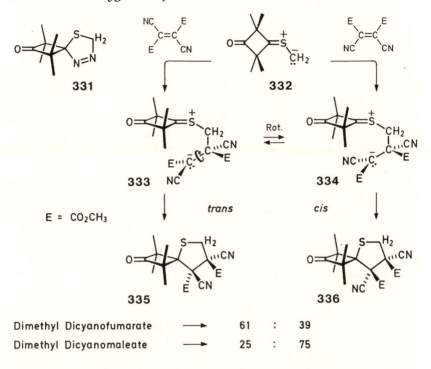

E = CO₂CH₃

| Dimethyl Dicyanofumarate | ⟶ | 61 | : | 39 |
| Dimethyl Dicyanomaleate | ⟶ | 25 | : | 75 |

Chart X. Two-step 1,3-cycloadditions of 2,2,4,4-tetramethyl-3-thioxocyclo-butanone S-methylide (332) to dimethyl 2,3-dicyanofumarate and dimethyl 2,3-dicyanomaleate (CDCl₃, 10 min at 85 °C).

Stabilization of the ketene imine was achieved when 2,3-bis-(trifluoromethyl)fumaronitrile served as acceptor olefin. The reaction with **332** at 40 °C provided the ketene imine **338** and the thiolane **340** in a 79:21 ratio (CDCl₃); **338** was obtained crystalline.[381a] This is a new example of a black-magic effect, the stabilization of strained molecules by fluorine or trifluoromethyl substitution. X-ray analysis confirmed

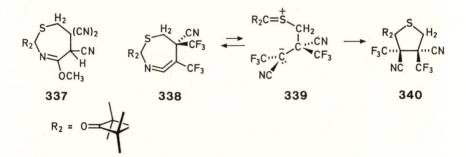

the structure of a seven-membered ketene imine closely related to 338;[381b] here as in other cases I enjoyed the neighborly help of Heinrich Nöth, my good colleague and friend.

How is 338 formed? The C–C cyclization of the *trans*-zwitterion 339 to give 340 is sterically hampered, and the competing N-alkylation by the sulfonium–carbenium ion leads to 338. At 60 °C, ketene imine 338 quantitatively rearranged to thiolane 340. Elke Langhals found this reaction 10^3 times faster in acetonitrile than in cyclohexane; that comparison provided good evidence for the intermediacy of the zwitterion 339.[381c]

For the Diels–Alder reaction, the crossing from the concerted to the stepwise mechanism is well documented.[382] The concurrence of activation enthalpy and entropy is important: high temperature, steric hindrance, and stabilization of the intermediate by substituents favor the two-step pathway. For many years, we were not able to divert 1,3-dipolar cycloadditions from the orthodox concerted pathway, and the final crossing of the borderline was a relief.

Historically, the clarification of reaction mechanisms has often been linked to unique model compounds. The aim is to draw *general* conclusions from the experience with a *specific* class of compounds. A good model should be representative, quite accessible, and analytically easy to handle. Great care should be given to preceding considerations and the final selection. In my experience, it is worthwhile to spend extra months on thinking and preliminary testing. The literature offers warning examples of poor choice of models. For example, α-halocarboxylic acids turned the clarification of the Walden inversion into a tragedy. The nucleophilic substitution of this class of compounds was fraught with complications.

Strained Cycloalkenes

Fast cycloadditions of phenyl azide served Alder and Stein[288] as a diagnostic tool for the double bond of bicyclo[2.2.1]heptene type (page 88). For 1,3-dipoles of the propargyl–allenyl type, we measured rate ratios up to 6100 in addition to norbornene and cyclohexene.[383]

The relief of angle strain in going from norbornene (342) to norbornane amounts to 4.7 kcal mol^{-1}; the cycloaddition enthalpy should be increased by roughly this energy. However, the expectation that the rates of cycloaddition to cycloalkenes would parallel the strain relief on hydrogenation was not fulfilled (Chart XI). A factor "x" must increase the reactivity of norbornene (342) compared with bicyclo[2.2.2]octene (341), bicyclo[2.1.1]hexene, and its derivative 343.[384] Pieter H. J. Ooms, a guest from Holland, did the kinetic measurements.

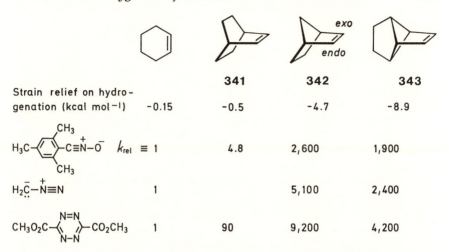

		341	342	343
Strain relief on hydrogenation (kcal mol^{-1})	-0.15	-0.5	-4.7	-8.9
(trimethylbenzonitrile N-oxide) $k_{rel} \equiv 1$		4.8	2,600	1,900
$H_2\bar{C}-\overset{+}{N}{\equiv}N$	1		5,100	2,400
$CH_3O_2C-\langle N{=}N / N{-}N \rangle-CO_2CH_3$	1		90	9,200	4,200

Chart XI. Strain relief on hydrogenation of cycloalkenes and relative rate constants of cycloadditions.

Not much imagination is required to connect "*x*" with the preferred *exo* reactivity of the less symmetric **342**. The *exo–endo* rate ratio versus diphenylnitrilimine is >300.[385] In apobornene (**344**), cycloaddition in *exo* has to overcome massive steric hindrance by 7-methyl. The addition of trimethylbenzonitrile *N*-oxide to **344** is 4300 times slower than that to **342** (CCl$_4$, 25 °C), but still displays an *exo–endo* ratio ≥70. If we assume identical rates for *endo* cycloadditions to **342** and **344**, the *exo* preference of norbornene should exceed 300,000, corresponding to $\Delta\Delta G^{\ddagger} \geq 7.5$ kcal mol^{-1}.[386] The predilection of norbornene for *exo* attack is also manifest in catalytic hydrogenation, permanganate oxidation, HCl addition, hydroboration, etc.

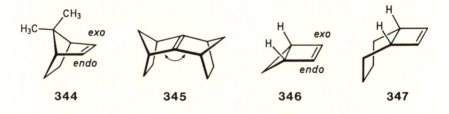

In 1967, Schleyer[387] ingeniously ascribed the *exo* preference in norbornene reactivity to a decrease of conformational strain in the transition state (TS); this torsional effect is worth ≤2 kcal mol^{-1}. In 1976 PMO calculations by Fukui et al.[388] resulted in nonequivalent orbital extension for the π bond of norbornene because of hyperconjugation.

Greater *exo* lobes in the ground state should be responsible for the *exo* preference.

Our factor "*x*" found attention and prompted quantum-chemical calculations. Wipff and Morokuma[389] found the olefinic C–H of norbornene *endo*-bent by 5°, but the π-electron density was hardly different for *exo* and *endo*. However, the interaction with a proton in 2-Å distance of the π bond is substantially more advantageous on the *exo* than on the *endo* side. According to calculations by Gleiter et al.,[390] out-of-plane *endo*-bending of the olefinic C–H in the TS costs less energy than *exo*-bending.

Organic chemists are accustomed to interpreting rate and energy data in terms of *effects*. However, the "black box" (computer) spews numbers only, and the theorist depends on dialog with the organic chemist for explanations in terms of effects. Gleiter et al.[390] ascribed the *endo* deformability in the TS to hyperconjugative interactions. Houk et al.[391] traced factor "*x*" back to a staggered arrangement of the allylic bonds with the partially formed new bonds in the TS. Doubts persist whether the calculated effects are high enough to explain both *exo* preference and the unusually high rate of norbornene reactions.

In 1981, X-ray analyses of derivatives of *syn*-sesquinorbornene (345) revealed *endo* deformations of the ethylenic σ plane by 16° and 18°.[392] A rich literature on this phenomenon has accumulated in the meantime.

A nonequivalent orbital extension was also calculated for bicyclo[2.1.0]pentene (346).[388] We confirmed the *exo* cycloadditions to 346 and found the rates of Diels–Alder reactions and 1,3-dipolar cycloadditions much higher than those of the cyclobutene 347. Rate ratios of 346–347 versus the following reagents were: tetraphenylcyclopentadienone, 52,000; tetrachloro-*o*-benzoquinone, 200,000; diphenyldiazomethane, 1700; and C-*p*-nitrobenzoyl-*N*-phenylnitrone, 2800, as measured by Adolf Nuber.[386] We doubt that a "Fukui effect" in the ground state of 346 can give rise to accelerations of this magnitude, particularly because ab initio calculated *endo*-bending of the olefinic C–H in 346 amounts to only 5°.[393]

Torsion and out-of-plane bending characterize the double bond of *trans*-cyclooctene (348) as partially broken (page 32).[68] In the rate constant of phenyl azide addition, 348 and *cis,trans*-1,5-cyclooctadiene (349) exceed norbornene 90-fold and 6000-fold, respectively.[394] The nitro group is highly stabilized by resonance. Leitich[395] was the first to observe 1,3-dipolar activity of the aromatic nitro group in thermal additions to 349 and to 348, the latter addition being 400 times slower. The two strained cycloalkenes likewise combined with aromatic azoxy compounds. The adduct 350 from 348 and 4,4'-dicyanoazoxybenzene harbors the weak N–O bond and experiences intramolecular 1,3-dipolar cycloreversion. The azomethine imine **351** tautomerized to the enehydrazine **352** (82%) or was trapped by an excess of **348** furnishing a 2:1 product.[396] This study was the more demanding, as the products were not crystalline; the careful work of Francisco Palacios, a guest from Oviedo (Spain), deserves recognition.

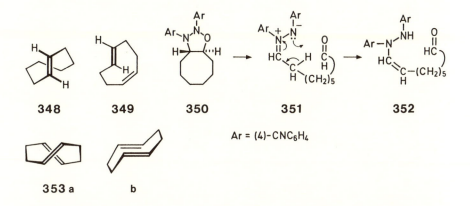

348   **349**   **350**   **351**   **352**

353 a   **b**

Ar = (4)-CNC$_6$H$_4$

In 1969 Whitesides, Cope et al.[397] obtained *trans,trans*-1,5-cyclooctadiene in 1% yield on irradiation of the CuCl-complexed *cis,cis* isomer. The constraint of the double bonds does not allow interconversion of the *racemic* and *meso* conformations, **353a** and **353b**. With skill and experience, Dieter Boeckh (Ph.D. 1986) converted *cis,cis*-1,5-cyclooctadiene via the bisepoxide into the very sensitive **353** in gram quantities and confirmed Cope's intuitive assumption of the crisscross structure **353a**.[398a] The structural elucidation of a precursor was an indirect argument for **353a**, and the X-ray analysis of the diazofluorene adduct provided direct evidence (H. Nöth).

The heat of formation (ΔH_f°) of cycloalkenes can be determined from heats of hydrogenation or by molecular mechanics (MM1). Comparison with ΔH_f° based on Benson's group increments[398b] provides the following strain energies: *cis*-cyclooctene, 5.9; *trans*-cyclooctene, 16.7; *cis,trans*-1,5-cyclooctadiene, 22.6; and *trans,trans*-cyclooctadiene, 31.8 kcal mol^{-1}.[394]

Electrocyclic Reactions

The scientific world of today is very alert to new ideas. The famous communications that Woodward and Hoffmann[351] published in 1965 on the principle of conservation of orbital symmetry quickly became the most frequently quoted papers in chemistry. The principle added a new color to our picture of the reaction event and stimulated a wealth of experimental and MO-theoretical work. Only 4 years after the discovery, an international symposium (in Cambridge in 1969) dealt with the new perspectives.

The Evans–Dewar–Zimmerman principle, an alternative approach to the selection rules, is founded on the topology of the basis set of atomic orbitals in the transition state. No longer symmetry, but rather Hückel- and Möbius-type aromaticity and antiaromaticity are the criteria.[399,400] In 1966 Fukui[401] proposed that reactions take place in the direction of maximum HO–LU overlap.

Electrocyclic reactions defined by Woodward and Hoffmann[402] are cyclizations of polyenes with concurrent shift of double and single bonds. The net result is the conversion of a π bond into a σ bond. Pertinent phenomena were known previously; they sailed under the flag of "valence tautomerizations". However, the field of electrocyclic reactions began to prosper only after the new ideas had opened our eyes. Orbital control imposes a definite steric course on the concerted bond reorganization, either conrotatory or disrotatory.

Valence Tautomerism of Cyclooctatetraene

Is cyclic conjugation a sufficient requirement for aromaticity? Studying cyclooctatetraene (COT), Willstätter and Waser[403] came to a negative answer; their 1911 synthesis of COT was fraught with difficulties. Thirty years later the catalytic tetramerization of acetylene allowed Reppe et al.[404] a closer investigation of the yellow COT, which became available in ton quantities. Whereas the products of hydrogenation and epoxidation are derived from the fourfold unsaturated eight-membered ring 354, the Diels–Alder adducts with dienophiles $d=e$, 356, and halogenation products like 357 are formally derived from the bicyclic 355.

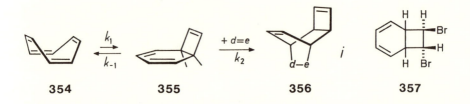

354   **355**   **356**   **357**

The basic work on COT was done at the BASF AG during World War II. Walter Reppe, a pioneer in acetylene chemistry and transition-metal catalysis, was director of research. When Reppe's now-classic 1948 paper[404] was submitted to *Liebigs Annalen*, Heinrich Wieland, the editor, asked me to serve as referee. The reviewer studies (or ought to study) a paper more carefully than the casual reader. There was a wealth of new chemistry in this milestone publication (92 pages). Several "reaction formulae" were used, such as 354 and 355, without clearly distinguishing between discrete molecules and skeletal rearrangements within the reaction complex.

The problem fascinated me, as did the similarly unsettled relation of cycloheptatriene and norcaradiene. In the late 1950s the alternatives—rearrangement 354 → 355 *before, during, or after* combination with the dienophile—had found advocates. The kinetic study carried out in the Munich laboratory presented an elegant experimental decision in favor of an electrocyclic equilibrium of COT by Fritz Mietzsch in 1964.[405]

Many chemists are unaware of the precision and beauty of dilatometric rate measurements. "Contractometric" would be a more fitting term because the combination of two molecules to give one is accompanied by volume shrinkage of the solution. The near-quantitative yields of Diels–Alder adducts 356 allowed dilatometry, which is only applicable to first-order or pseudo-first-order reactions.

The combinations of COT with large and variable excess concentrations of dienophiles D furnished dilatometric rate constants, k_d. Steady-state treatment of the system with a reversibly formed isomer of COT entering into the Diels–Alder reaction with D led to eq 2.

$$k_d = \frac{k_1 k_2 D}{k_{-1} + k_2 D} \tag{2}$$

Less reactive dienophiles like maleic anhydride fully appeared with first order in the rate equation; $k_{-1} >> k_2 D$ simplifies eq 2 to $k_d = k_1 k_2 D / k_{-1}$. On using increasingly reactive dienophiles [i.e., tetracyanoethylene (TCNE) and dicyanomaleimide], the tautomeric equilibrium is disturbed; $k_2 D$ competes with the reverse reaction (k_{-1}). The plot of k_d versus the dienophile concentration afforded curves that approached a plateau value corresponding to $k_d = k_1$; the evaluation yielded the tautomerization constant k_1 and the ratio k_{-1}/k_2. Finally, when the Diels–Alder step is very fast, the tautomerization with k_1 determines the rate alone. For phenylcyclooctatetraene + dicyanomaleimide, the dilatometric rate constants were independent of the dienophile concentration ($k_d = k_1$).

Dilatometry, as it became a tradition in our laboratory, required skill and experience. The co-workers joked that after 6 weeks either the technique is mastered or there will be a case for the mental hospital. Fortunately, none of them ended up in the clinic.

The kinetic data left no doubt that the bimolecular Diels–Alder reaction of COT is preceded by a first-order equilibration;[405] this is accepted textbook knowledge by now. The tetraene 354 cannot directly combine with the dienophile because the double bonds in the boat conformation (electron diffraction data[406]) are fixed in a close-to-orthogonal arrangement; a nearly planar diene system is prerequisite to the Diels–Alder reaction. Bicyclo[4.2.0]octatriene (355) harbors such a planar 1,3-diene system and is a plausible intermediate. Support for 355 came from the debromination of dibromide 357 at −78 °C, as studied by Vogel, Kiefer, and Roth;[407] 355 rearranged to 354 at 0 °C.

Our kinetic data established a barrier of $\Delta H^{\ddagger} = 27$ kcal mol^{-1} and $\Delta S^{\ddagger} = -1$ eu for the electrocyclization 354 → 355. Knowledge of k_{-1} would provide the equilibrium constant, but only k_{-1}/k_2 was available. A trick helped. Structure 356 discloses that the dienophile $d=e$ approaches 355 from the less hindered side, thus backing up the assumption of similar k_2 values of 355 and bicyclo[4.2.0]octadiene (363). The latter is isolable and its cycloaddition rate was measured. Now an approximate value for k_{-1} was calculated, and an equilibrium concentration as small as 0.01% 355 (dioxane, 100 °C) resulted, corresponding to $\Delta G = +7.3$ kcal mol^{-1} for 354 → 355.[408]

In 1986, Squillacote and Bergman[409] established the equilibrium 354 ⇌ 355 in the gas phase at 400–700 °C and 10^{-5} torr; ^{1}H NMR analysis of 355 in the frozen samples furnished ΔG (100 °C) = 7.1 kcal mol^{-1}, in excellent agreement with our indirectly obtained value.

We interpreted the tautomerization **354** → **355** by a cyclic electron shift of the type $3\pi \rightleftarrows 2\pi + \sigma$. The likewise conceivable shift $2\pi \rightleftarrows \pi + \sigma$ was considered "less probable after examination of models" in 1964. One year later, Woodward and Hoffmann[402] pointed out that only the hexatriene system in **354** was allowed a *disrotatory* cyclization by orbital symmetry.

We applied the same kinetic method to establish intermediates in cycloadditions to related unsaturated systems. A significant failure preceded our work with COT: The equilibrium of cycloheptatriene (**358**) with norcaradiene (**359**)[410] is so rapidly established that cycloadditions with potent dienophiles $d=e$ (TCNE, perfluorocyclobutanone, and *N*-phenyl-1,2,4-triazoline-3,5-dione) yielding adducts **360** remained the slow step.[411,412] David England, a guest of the DuPont Company, and Werner Hentschel as a graduate student were "tragic heroes" because their hard work did not allow binding conclusions. The strategem of borrowing k_2 from **363** would be in harmony with an equilibrium concentration of 0.15% of norcaradiene (**359**) along with **358** at 25 °C.[412] We are facing here the fastest electrocyclic reaction known, probably because of *homoaromatic stabilization* of the transition state.

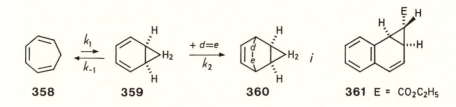

358   **359**   **360**   **361** E = $CO_2C_2H_5$

In 1981, the parent norcaradiene (**359**) was seen by Mordecai Rubin.[413] Generated by a photoreaction in a hydrocarbon glass at 93–103 K, the electrocyclic reaction **359** → **358** advanced via $\Delta H^{\ddagger} = 6.5$ kcal mol^{-1}; possibly the barrier is smaller in a more fluid medium.

The incorporation of a double bond of norcaradiene into an aromatic ring should increase the activation barrier of valence tautomerization and shift the equilibrium to the benzonorcaradiene side in **361**. Indeed, the *o*-quinodimethane-type cycloheptatriene analog was not observed by Gottfried Juppe up to 200 °C.[414]

Cycloocta-1,3,5-triene and Derivatives

In a pioneering study, Cope et al.[415] reported in 1952 on the valence-tautomeric equilibrium **362** $\rightleftarrows$ **363** (85:15 at 100 °C). The electrocyclic reaction proved so slow at room temperature that triene **362** could be separated as AgNO₃ complex. Bicyclo[4.2.0]octadiene (**363**), but not the monocyclic triene **362**, combined with maleic anhydride at 10 °C to give **364**.

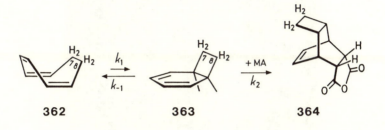

On reacting pure **362** with excess concentration of maleic anhydride at 85–100 °C, Gernot Boche observed the curvature in the plot of k_d versus D that is typical for $k_{-1} \sim k_2 D$ in eq 2. The plateau value, $k_d = k_1$, was reached in dilatometric experiments with all excess concentrations of TCNE. The pure bicyclic tautomer **363** was available here, and all the rate constants and Eyring parameters were measured for this system.[408] Interestingly, the electrocyclization barriers for COT (**354**) and cyclooctatriene (**362**) were nearly identical, but ΔG (100 °C) values of 1.4 kcal mol^{-1} for **362** → **363** and 7.3 kcal mol^{-1} for **354** → **355** reflect the higher ring strain of the cyclobutene derivative **355** compared to cyclobutane **363**.

The influence of substituents at the saturated centers 7 and 8 of **362** on the electrocyclic equilibrium was less clear. The percentage of the bicyclooctadiene form at 60 °C rose from 11% for the parent **363** to 53% for the 7-acetoxy, 94% for *trans*-7,8-dimethyl, and 99% for the *trans*-7,8-dichloro derivative.[416]

Rearrangement of Bromocyclooctatetraene

The nonaromatic 8π system of COT and its derivatives unerringly finds its way to 6π aromaticity by imaginative wiggles. COT is a true molecular acrobat, and I became quite fond of that simple molecule.

Cope and Burg[417] described the conversion of bromocyclooctatetraene (**365**) to β-bromostyrene (**371**) at 100 °C. This rearrangement constitutes the kind of puzzle organic chemists like. Will Elmar Konz (Ph.D. 1970) did a marvelous job of fitting the pieces together.

We observed a quantitative conversion of **365** at 80 °C to *trans*-β-bromostyrene with 99.9% stereoselectivity; no intermediate became visi-

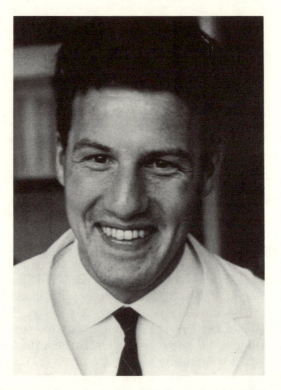

Gernot Boche as a graduate student in 1963; he received his Ph.D. in 1965. His habilitation thesis (Munich 1974) dealt with electrocyclic reactions and [9]annulene anions. A professor in Marburg since 1979, he successfully studies reactivity and structure of organolithium compounds.

ble by NMR spectroscopy during the reaction. The overall first-order rate constant increased with growing solvent polarity by a factor of 600 until—suddenly—a plateau was reached for acetonitrile, acetic acid, and methanol. For the less polar solvents, log k was a linear function of the empirical parameter of solvent polarity, E_T. Lithium iodide did not interact with **365** or **371**, but interfered during the rearrangement; *trans-β*-iodostyrene appeared. The acid catalysis likewise supported an ionization step as rate-determining until—at the plateau—a preceding step that was not influenced by solvent polarity took over the rate control.[418]

The sequence with electrocyclization of **365** giving 1-bromocyclooctatriene (**366**), ionization to the homocyclopropenium salt **367**, ion recombination on the upper side, and conrotatory opening of the cyclobutene ring of **368** was attractive. The rate plateau in polar solvents is the disrotatory cyclization **365** → **366**. Bromine in **371** is no longer

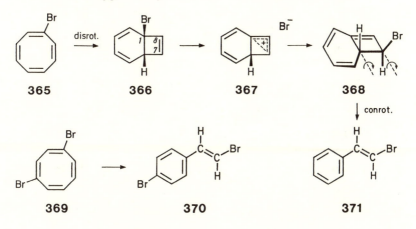

365   366   367   368

369   370   371

attached to the original carbon atom but has undergone a 1,3 migration. This migration was confirmed by the rearrangement of 1,4-dibromocyclooctatetraene (**369**) giving 92% of **370** containing the bromines in 1,6 positions.[419]

There was still a serious obstacle to overcome. The Diels–Alder reaction of **365** with TCNE yielded only the adduct that originated from 7-bromobicyclooctatriene. The latter simply turned out to react with preference. N-Phenyl-1,2,4-triazolinedione, as the more potent dienophile, also trapped the 1-bromo isomer **366** to the extent of 25%.[420]

Halogenation of Cyclooctatetraene

Reppe et al.[404] reported the bicyclic *trans*-dichloride **375** as a chlorination product of COT, and the analogous structure **357** was assigned to the dibromide. We found that the bromination of COT is still fast at −55 °C; the bicyclic tautomer **355** cannot be an intermediate, because the electrocyclization **354** → **355** would require a half-reaction time of 48 million years at −55 °C.

In 1964, a Varian A60 instrument opened the NMR era in the Munich Laboratory, and the elucidation of the halogenation mechanism profited from the new tool. More than 90% of *cis*-7,8-dichlorocyclooctatriene (**372**) was isolated after chlorination of COT at −40 °C, and above 50 °C electrocyclic equilibrium with **373** was attained. Alumina or acid as catalysts established a *cis–trans* equilibrium (34:66 at −30 °C) of the monocyclic dichlorides, **372** ⇄ **374**; the electrocyclization gave **375** in the final step. All four dichlorides of COT, **372–375**, were obtained crystalline and their structures were clarified by Wolfgang Hechtl.[421]

Bromination of COT took an analogous course, likewise starting

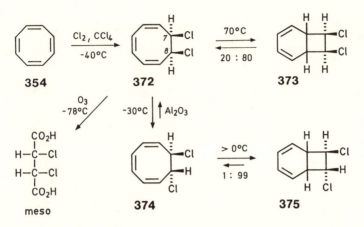

with a selective *cis*-addition.[422] Treatment of the *cis*-dichloride 372 or the *cis*-dibromide by potassium *tert*-butoxide at −45 °C made chloro- and bromocyclooctatetraene—key substances in COT chemistry—easily accessible for the first time. I received numerous inquiries for the detailed procedure before our full paper was published in 1971.[423] This finding was a contribution by our Australian guest, George E. Gream, University of Adelaide; after his return he continued to devote his research to COT chemistry.

The high rate and the *cis*-selectivity of the initial halogen addition posed tough problems. More than 100 years ago, Kekulé sketched a picture of the bromination of ethylene in which the breaking of the Br−Br bond and the formation of both C−Br bonds occur simultaneously. Orbital control forbids a *cis*-bromination to be concerted. On the other hand, the favored *cis*-1,4-bromination of cyclopentadiene may very well be a one-step reaction. Also, the concerted *cis*-1,8-halogenation of COT via 376 would be allowed. A counterargument is as follows: Conjugation of the double bonds in the boat conformation 354 of COT is sterically hindered (i.e., the 8π electrons are not "socially integrated"). The problem bothered me for many months, and I felt relieved when the solution was found. Once the idea was born that an 8-halohomotropylium ion might be the culprit, its occurrence was easily demonstrated.

Winstein's fruitful concept of homoaromaticity[424] was much discussed in the 1960s. In 1966 Winstein et al.[425] described the conversion of COT by deuterated concentrated sulfuric acid into 8-*endo*- and 8-*exo*-deuteriohomotropylium ion in an 80:20 ratio.

Fluorosulfonic acid converted our *cis*-dichloride 372 in liquid SO_2 at −20 °C into the *endo*-8-chlorohomotropylium salt 377 (FSO_3^- instead of Cl⁻), which undergoes ring inversion at 20 °C, affording the more stable *exo* isomer 378 (FSO_3^-).[426] The latter was likewise obtained from

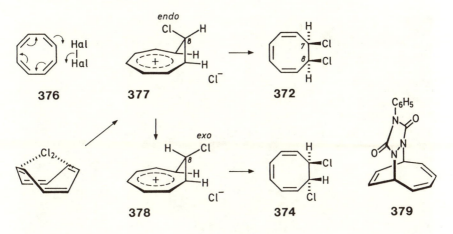

the *trans*-dichloride **372** by FSO$_3$H. The difference of δ_H (8-*endo*) and (8-*exo*) amounted to 5.7 ppm, a powerful argument for the presence of a homoaromatic ring current. The two 8-chlorohomotropylium ions were shown to accept Cl$^-$ from the *endo* side, yielding **372** and **374**, respectively.[427] These demanding experiments required the skill of Gernot Boche.

How can the initiating step be established? Treatment of COT with 2 equiv of SbCl$_5$ at −60 °C provided pure *endo*-8-chlorohomotropylium hexachloroantimonate (**377**, SbCl$_6^-$ instead of Cl$^-$); SbCl$_5$ plays a unique double role as a Cl$^+$ donor and complexing reagent. This key experiment by Johann Gasteiger (Ph.D. 1971) strongly suggested **377** as primary product of the chlorination.[428] In 1977 Olah et al.[429] used Sb(V) as an oxidant to convert methylated and phenylated cyclooctatetraenes into Hückel-type dications.

The halogenation of COT turned out to be a drama in many acts. Elucidation of a reaction mechanism has no more than a relative meaning. The *perpetuum mobile* of solved problems generating new ones goes into the next round. Why does Cl$^+$ attack COT from the *endo* side? Why is *exo*-8-chlorohomotropylium ion (**378**) energetically favored over the *endo* isomer **377**? Why do both ions combine with Cl$^-$ on the *endo* side? Why is the electrocyclic reaction **374** ⇌ **375** so unusually fast?

Walter Reppe was a powerful personality. At first contact appearing somewhat abrasive, he exemplified the soft core in a hard shell. In 1967 I told Reppe—by then retired—about the mechanism of formation of Reppe's dichloride **375**. The pioneer's arid remark, "You better find an application and make money", sheds light on the difficulties of doing fundamental research in industry without the bonus of successful application. Cost is a merciless judge in the industrial utilization of beautiful discoveries. Reppe's office at the BASF was furnished

Walter Reppe (1892–1969), BASF AG, was one of the most distinguished chemists in German industrial research. He pioneered in acetylene reactions under pressure and made acetylene a major building block in the production of organic chemicals. (Photograph courtesy of BASF AG.)

with a colorful rug bearing a pattern of COT formulae and woven from polycaprylolactam (nylon 8) fiber, the latter made from COT. Reppe liked to puzzle visitors by admonishing them, "Don't step on my COT molecules."

Interaction of π bonds allows conjugation or benzene-type aromaticity. COT escapes the *antiaromaticity* by steric hindrance of resonance in the boat form **354**, but strives for a more favorable number of π electrons. The acceptance of two electrons produces the planar eight-membered Hückel–aromatic dianion, whereas good electrophiles procure an energy gain by converting COT to homotropylium ions.

More Homotropylium Intermediates and Electrocyclizations

N-Phenyl-1,2,4-triazolinedione reacted as a dienophile with the bicyclic tautomer **355** to give a Diels–Alder adduct of type **356**. As a potent electrophile, it also attacked the monocyclic COT, providing a homotropylium zwitterion that collapsed to the formal 1,4-cycloadduct **379**.[430] When the nucleophilicity of COT is increased by a methoxy or phenoxy group, even TCNE chose the homotropylium route affording 1,2- and 1,4-cycloadducts,[431] as revealed by Johann Gasteiger.

COT is stable in sulfur dioxide, but in the presence of 1 equiv of SbF$_5$ the sulfone **380** was formed as a 1,4-adduct via the homotropylium species **381**. A second equivalent of SbF$_5$ shifted the energy balance in favor of the open-chain zwitterion **382**.[432]

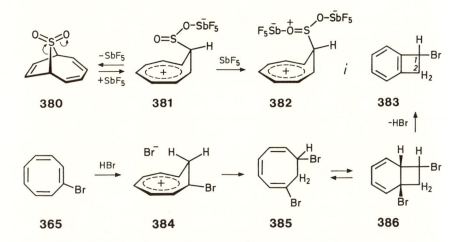

Treatment of bromocyclooctatetraene (**365**) with hydrogen bromide in acetic acid at 60 °C furnished 1-bromobenzocyclobutene (**383**). The 1-bromohomotropylium ion **384** combines with Br⁻ to give the dibromocyclooctatriene (**385**). The electrocyclic equilibrium, **385** ⇌ **386**, is disturbed by aromatization achieved through formal *cis* elimination of HBr. This and other delightful rearrangements belonged to the late harvest brought in by Ulrich Schnegg.[433]

Benzocyclobutenes and *o*-Quinodimethanes

Several authors have observed cycloadditions of benzocyclobutenes to electron-deficient double bonds, and in 1958 Jensen and Coleman[434] delineated the mechanistic alternatives: Either a ring opening to an *o*-quinodimethane precedes a Diels–Alder reaction, or the opening of the four-membered ring is concerted with the attack by the alkene. My co-worker Helmut Seidl and I were the first to establish an intermediate and the steric course of the bond reorganization.

In 1964 dilatometric rate measurements of the reactions of *trans*- and *cis*-1,2-diphenylbenzocyclobutene (**387** and **390**) with TCNE at 50 °C proved the occurrence of an intermediate: k_d was *independent* of the TCNE concentration (i.e., the electrocyclic ring opening with k_1 alone controlled the rate). The curves observed when k_d was plotted versus the concentration of maleic anhydride demonstrated $k_{-1} \sim k_2 D$ in the framework of eq 2 (page 127).[435]

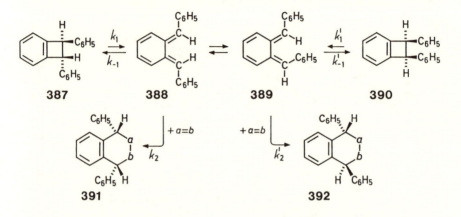

The quinodimethanes **388** and **389** are the intermediates of choice; they gave rise to different Diels–Alder adducts. The NMR spectra of the maleic anhydride adducts revealed the configurations. The aromatic protons appear as an AA'BB' pattern in the adduct **391** and as an ABCD spectrum in **392**. Thus, the electrocyclic ring opening took place by a steric course defined as *conrotatory* 1 year later (1965) by Woodward and Hoffmann.[402] Two preceding observations showed the same conrotatory conversion of monocyclic cyclobutenes to butadienes.[436,437] What a surprise that the incorporation of the double bond of cyclobutene into the benzene ring of **387** and **390** did not invalidate the rules of conservation of orbital symmetry. In the absence of dienophiles, a slow stereoisomerization led to a 92:8 equilibrium of **387** and **390**.[435]

The same model **387** and **390** was used by Quinkert et al.[438] in 1969 to demonstrate nonstereospecific photolysis of the four-membered ring at −185 °C. Recently, Takahashi and Kochi[439] generated the radical cations of **387** and **390** by irradiation in the presence of TCNE or 9,10-dicyanoanthracene; conrotatory ring opening and subsequent rotation are much faster in the radical cations than in the neutral species.

In the 1970s intramolecular Diels–Alder reactions of benzocyclobutene derivatives via *o*-quinodimethanes gained importance in the stereoselective synthesis of natural products; Oppolzer synthesized alkaloids, and Kametani steroids (estrone, estradiol).[173,302,303] Likewise in the 1970s, the parent *o*-quinodimethane was investigated by matrix isolation techniques.[440]

The stream of publications has not ceased yet, but—30 years later—our pioneer study is hardly quoted anymore. When a reaction or phenomenon, upon discovery, quickly becomes general knowledge, quoting may be dispensable. Scientists come and go, but science (hopefully) remains. However, it is less excusable to quote a casual later application by a peer and to disregard the original discoverer.

Deca-2,4,6,8-tetraenes

The intracyclic disrotatory ring closure of cycloocta-1,3,5-triene (362) was described earlier. Compound 362 itself could be the product of an electrocyclization of *cis,cis*-2,4,6,8-octatetraene. Woodward and Hoffmann[402] had predicted conrotation for this process, and Alexander Dahmen, Helmut Huber, and I reported the verification in 1967.[441] Enthusiasm for the predictive power of the conservation rules, a novel feature of MO theory, stimulated our stereochemical study.

Terminal methyl groups served as stereochemical markers. The preparation of the three isomeric decatetraenes 393, 396, and 397 in the pure and crystalline state was no trifle. Many months were required to elaborate conditions for the partial hydrogenation of the deca-2,8-

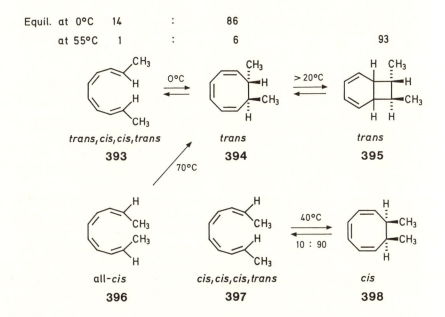

diene-4,6-diynes: Lindlar's palladium catalyst in pentane at −20 °C and stopping after uptake of 2.4 mol of H_2.

At 0 °C the *trans,cis,cis,trans*-isomer **393** cyclized by conrotation to *trans*-7,8-dimethylcyclooctatriene (**394**) until a 14:86 equilibrium was attained. At 50 °C, the next electrocyclic step, the disrotatory reaction **394** ⇄ **395** became mobile. The cyclization of the all-*cis*-decatetraene (**396**) at 70 °C furnished the same *trans*-dimethyl system, **394** ⇄ **395**; **396** was not discernible after equilibration. Conrotatory cyclization of the *cis,cis,cis,trans*-isomer **397** occurred at 40 °C and provided *cis*-7,8-dimeth-ylcyclooctatriene (**398**). The equilibrium, fully established at 55 °C, en-compassed 2% of **397**, 17% of **398**, and 81% of *cis*-7,8-dimethylbicy-clo[4.2.0]octadiene.[441]

The electrocyclization **393** → **394** is the fastest ever observed for an open-chain polyene. The cyclization of the lower vinylog, *trans,-cis,trans*-octatriene (**399**),[442] is 12 billion times slower at 0 °C (Chart XII). Is it not a paradox that an eight-membered ring should be closed so much faster from an open-chain polyene than an ordinary six-membered ring? The terminal olefinic C atoms of **393** are close to each other in the helical conformation **393A**, and conrotatory cyclization profits immedi-ately from the incipient formation of the σ bond.[443] In contrast, the dis-rotation of **399** requires strong twisting of the triene system before some σ bond energy is gained from overlap of the terminal orbitals.

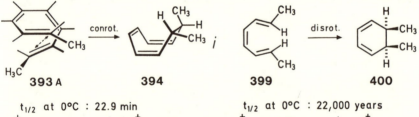

$t_{1/2}$ at 0°C : 22.9 min

$\Delta H^{\ddagger}$ = 15.1 kcal mol^{-1}, $\Delta S^{\ddagger}$ = −19 e.u.

$t_{1/2}$ at 0°C : 22,000 years

$\Delta H^{\ddagger}$ = 29.4 kcal mol^{-1}, $\Delta S^{\ddagger}$ = −7 e.u.

Chart XII. Half-reaction times and Eyring parameters for electrocycliza-tions of trans,cis,cis,trans-*decatetraene and* trans,cis,trans-*octatriene at 0 °C.*

Models show that one or two *cis*-methyl groups collide with the helix coil of decatetraene and render the overlap of the terminal π orbi-tals more difficult. The activation free energy of the ring closure is 5 kcal mol^{-1} higher for all-*cis*-decatetraene **396** than for **393**, and **397** lies in between.[443]

What does orbital-symmetry-allowed and -forbidden mean in terms of activation energy? Our decatetraene system permitted the first quantitative determination of such an energy difference for an electrocyclic reaction. A combination of rate and equilibrium measurements by gas chromatography disclosed that **393** at 171 °C undergoes one disrotation per 625,000 conrotations, corresponding to $\Delta\Delta G^{\ddagger} = 12$ kcal mol^{-1}.[444] Alexander Dahmen deserves special praise for his beautiful experimental work (Ph.D. 1969).

The nature of the forbidden reaction has not been unveiled here or in any other electrocyclic reaction. Although the ground state of the reactant is symmetry-correlated with a higher excited state of the product, *configuration interaction* allows reaching the ground state of the product, but via an energy barrier. A second mechanistic possibility lies in a slow *trans–cis* isomerization of a terminal double bond of **393**.[444]

Amusingly, the electrocyclization of the octatetraene system appears to occur in nature. Endiandric acid A (**403**) with its eight stereocenters was isolated from an Australian plant as a *racemate*. D. St. C. Black et al.[445] proposed that the polyunsaturated acid **401** undergoes in vivo the conrotatory octatetraene ring closure, followed by disrotatory hexatriene cyclization of **402** and a concluding intramolecular Diels–Alder reaction. In 1982, Nicolaou et al.[446] elegantly verified the sequence **401** → **403** with the methyl ester in a one-pot procedure.

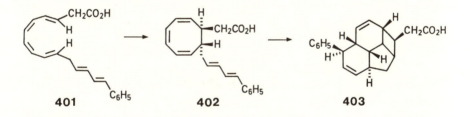

Aziridines and Azomethine Ylides

The Woodward–Hoffmann rules included polyenyl species.[402] The prediction of conrotatory ring opening for the conversion of cyclopropyl to allyl anions was all the more daring, as no clear-cut example was known in 1965. Today the reaction is well established, but the evidence for conrotation is still unsatisfactory. What is the difficulty? Substituted allyl anions are capable of rotation. In order to inspect their original structure, they must be transformed quickly to stable derivatives with retention of configuration. 1,3-Cycloadditions to phenylated ethylenes have been described, but their stereospecificity and concerted nature are in doubt.[447] Furthermore, rotation in allyl anions appears to be faster than their interception by cycloaddition, thus thwarting conclusions on the mode of ring opening of cyclopropyl anions. This phenomenon was demonstrated by my former associate Gernot Boche.[448]

The allyl anion is the electronic prototype of 1,3-dipoles, but it lacks the right blend of nucleophilic and electrophilic character. Replacement of the middle C–H by N–R (i.e., an iminium function) gives rise to azomethine ylides (Chart V), which are isoelectronic with allyl anions. The same relation holds for aziridines and cyclopropyl anions. Substituents that stabilize negative charge should activate aziridines for this hitherto unknown ring cleavage at the C–C bond.

A triple observation in 1965–1966 concerned cycloadditions of substituted aziridines to C–C double and triple bonds.[449–451] The reactivity was ascribed in the Munich report[451] to a small concentration of an azomethine ylide occurring in equilibrium with the aziridine. In 1967 we announced the first confirmation of the ring-opening modes predicted for the isoelectronic allyl anion.[452]

Thermal *cis–trans* equilibration of the dimethyl aziridine-2,3-dicarboxylates **404** and **407** at 100 °C proceeded via the open-chain azomethine ylides **405** and **408**. Their interception by active dipolarophiles like dimethyl acetylenedicarboxylate suppressed the *cis–trans* isomerization, and stereospecific 1,3-dipolar cycloadditions were observed. The structures of **406** and **409** revealed *thermal conrotation* and *photochemical disrotation* for the electrocyclic processes, a result of even aesthetic appeal.[452] The *exo,exo* configuration of the *cis*-diester **408** was deduced from NMR spectroscopic comparisons of further cycloadducts.[453]

Wolfgang Scheer had magic hands in experimenting. I did not then object to the beer bottles on his bench. Munich has the highest per capita consumption of beer in the world, and even the Roman historian Tacitus (55–116 A.D.) blamed the German tribes for drinking too much of the stuff. The prohibition of beer at Munich workplaces would be regarded as unpatriotic; beer can even be purchased in our institute's cafeteria. As for azomethine ylide chemistry, the very good work by Hansjoachim Mäder, J. Herbert Hall (1931–1990), Carl Heinz Ross, and Karl Niklas may likewise be acknowledged.

Our dilatometers and eq 2 (page 127) proved useful again. The rate constants k_d for the reactions of aziridines **404** and **407** with TCNE or diethyl fumarate were independent of the nature and concentration of the trapping reagents, thus establishing the electrocyclic ring opening as the slow step (i.e., $k_d = k_1$).[454,455] Flash spectroscopy afforded half-reaction times of 5.4 and 7.8 s at 25 °C for the recyclization of the yellow azomethine ylides, **405** → **404** and **408** → **407**.[456] An ensemble of kinetic data provided the complete energy profile of interconversions (Figure 3).[457]

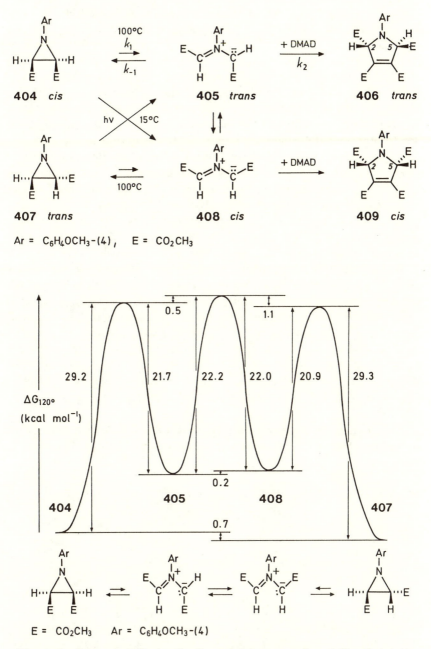

Figure 3. Energy profile for the thermal interconversion of dimethyl 1-(4-methoxyphenyl)aziridine-2,3-cis- and -trans-dicarboxylate.

The energy level of the azomethine ylides **405** and **408** lies 8 kcal mol^{-1} above that of the cyclic tautomers. The recyclization requires two 90° rotations about the C–N bond axes; the resonance energy of the 1,3-dipole is lost before the incipient σ bonding contributes much. This energy loss explains the astonishingly high recyclization barriers (21.7 and 20.9 kcal mol^{-1} in Figure 3). 1,3-Dipolar cycloadditions of the azomethine ylides **405** and **408** are free of this disadvantage. In contrast to the usual rate preference of intramolecular processes, intermolecular cycloaddition wins the competition over electrocyclic ring closure.

As a very active 1,3-dipole, azomethine ylide **405** adds even to aromatic bonds. The 1,2 additions to naphthalene or anthracene were followed by faster additions to the styrene-type double bonds of the monoadducts, and bisadducts were obtained.[458]

The *cis-* and *trans-*1,2,3-triphenylaziridines (**410** and **413**), our second model, required 150 °C for equilibration (79:21), but cycloadditions took place at 100 °C. The energy profile revealed that the rotational barriers of the azomethine ylides (27 and 28 kcal mol^{-1}) exceeded those of recyclization by 8 and 10 kcal mol^{-1}.[459] Thus, aziridines **410** and **413** were the models of choice for studying the *retention of 1,3-dipole configuration* in cycloadditions. Among the 1,3-dipoles of Chart V, only azomethine ylides and carbonyl ylides possess *trigonal carbon atoms* as termini. The reactions of **410** and **413** with tetraethyl ethylenetetracarboxylate proceeded quantitatively and with a stereospecificity of >99% (1% analytical limit). The stereochemical evidence refers to conrotatory ring openings *and* 1,3-cycloadditions of azomethine ylides **411** and **412**.[459,460]

The half-life of azomethine ylides **411** and **412** amounts only to 12 and 10 ms at 25 °C.[459] However, stabilization by substituents is feasible,

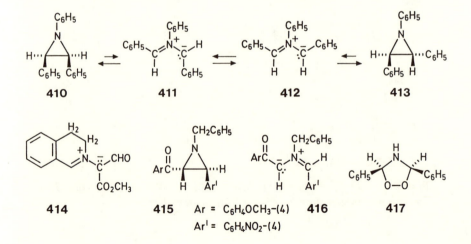

410   411   412   413

414   415   Ar = C$_6$H$_4$OCH$_3$-(4)   416   417
            Ar^I = C$_6$H$_4$NO$_2$-(4)

and the orange **414** was the first crystalline azomethine ylide;[461] it refuses to cyclize, but preserves some 1,3-dipolar activity.[462]

DoMinh and Trozzolo[463] discovered the photochromism of 2-aroyl-3-arylaziridines in the solid state; for example, the blue color of **416** developed on UV irradiation of **415** and was stable at 25 °C for 15 min in the dark, whereas visible light effected recyclization (**416** → **415**). Some bicyclic aziridines on filter paper produced red and blue colors lasting for days.[464] In 1975, on the eve of the National Symposium at Fort Collins, where I received the Roger Adams Award, Tony Trozzolo practiced magic; after a short UV exposure, the writing appeared in bright color: "The ylides are greeting you, Dr. Huisgen."

In Munich and in numerous other laboratories the synthetic potential of azomethine ylides was exploited.[465] According to Schaap et al.,[466] photooxygenation of both *cis*- and *trans*-2,3-diphenylaziridine in the presence of 9,10-dicyanoanthracene (DCA) as a charge-transfer sensi-

Receiving the Roger Adams Award in 1975. The Department of Chemistry at Colorado State University, Fort Collins, wanted to present something of lasting value to me, found itself in a predicament because the U.S. dollar was not the strongest at the time. Unfortunately, there is only one way to take a Stetson home. "Hey, where is your horse?" I was asked in New York on my way back.

tizer produced *cis*-3,5-diphenyl-1,2,4-dioxazolidine (**417**). The radical cation of the azomethine ylide involved rotates fast and assumes the stable
exo,exo conformation before reaccepting the electron from $O_2^{\bullet-}$; supposedly, **417** comes from the neutral azomethine ylide + singlet oxygen.
On DCA-sensitized irradiation, Laurent, Czebulska, et al.[467] observed
nonstereospecific cycloadditions of 1-butyl-*cis*-2,3-diphenylaziridine to
electron-deficient dipolarophiles.

Oxiranes and Carbonyl Ylides

Oxiranes are likewise isoelectronic with the cyclopropyl anion. In 1965
Linn and Benson[468] observed that at 120 °C tetracyanoethylene oxide
(**418**) adds to C–C double and triple bonds. A kinetic study by Linn[469]
revealed that the cycloaddition to styrene is preceded by a first-order
conversion to an "activated species", which is **419**.

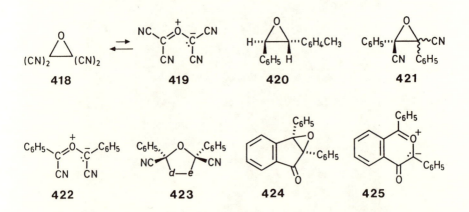

The comparison of ring-opening rates of aziridine **413** (half-life 11
min at 70 °C)[459] and oxirane **420** (218 min at 207 °C)[470] leaves no doubt
that carbonyl ylides are energetically less favored than azomethine
ylides; carbonyl compounds are weaker bases than imines. According to
ab initio calculations by Hehre,[471] the rotational barriers of the parent
azomethine ylide amounts to 29 kcal mol^{-1} and that of the carbonyl
ylide to only 14 kcal mol.$^{-1}$

Our first model, α,β-dicyano-*trans*- and -*cis*-stilbene oxides (**421**)
smoothly combined at 100 °C with electron-rich and electron-deficient
alkenes and alkynes, but—because of rotation—failed to inform on the
steric course of ring opening; as found by Helmut Hamberger, both *cis*-
and *trans*-**421** furnished the same cycloadducts **423** derived from **422**.[472]

In the thermal equilibration of the tricyclic oxirane **424** with the red carbonyl ylide **425** (isochromylium 4-olate), observed by Ullman and Milks,[473] the symmetry-forbidden disrotation is forced by the cyclic array.

α-Cyanostilbene oxides (**426** and **430**) turned out to be superior models. After initial ventures by Alexander Dahmen, the efforts of Volker Markowski (Ph.D. 1974) gave deeper insight. In 1971 we reported[474] the quantitative formation of the tetrahydrofurans **428** and **429** (*cis* with respect to phenyls) from α-cyano-*trans*-stilbene oxide (**426**) and dimethyl fumarate at 130 °C, thus confirming the predicted thermal conrotation for the electrocyclic reaction **426** ⇌ **427**. Dilatometric rate measurements disclosed a competition of recyclization (k_{-1} and cycload-dition (k_2D); $\Delta H^{\ddagger} = 30$ kcal mol^{-1} resulted for **426** → **427**.[475]

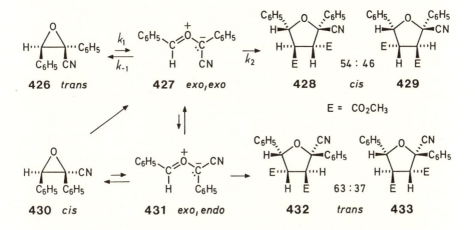

E = CO$_2$CH$_3$

The ring-opening rate of α-cyano-*cis*-stilbene oxide (**430**) at 130 °C was slower than that of **426** by a factor of 150. This difference was ascribed to the higher energy level of carbonyl ylide **431**, which was caused by the twisted *endo*-phenyl.[475] Intriguingly, in molten dimethyl fumarate at 140 °C the *cis*-oxirane **430** furnished 54% of the *cis*-adducts **428** + **429** in addition to 41% of the expected *trans*-adducts **432** + **433**.[474]

The ratio of *cis*- and *trans*-cycloadducts obtained from **430** depended on the concentration of dimethyl fumarate in chlorobenzene. It came as a surprise that two reactions unquestionably contributed to the nonstereospecific portion. Only 64% of the **430** reacted via the *exo,endo* carbonyl ylide **431** and was subjected to the competition of bimolecular cycloaddition (→ **432** + **433**) and rotation to the favored *exo,exo* carbonyl ylide **427**. The other 36% of the *cis*-oxirane **430** reached **427** on a path shunning **431**.[476] Among several mechanisms, the forbidden disrotatory but sterically more advantageous ring opening **430** → **427** was preferred.

The free energy profile for the interconversion of **426** and **430** via carbonyl ylides **427** and **431** was determinable from kinetic data, but not with the same certainty as in the aziridine isomerizations via azomethine ylides (Figure 3).[477] The carbonyl ylides **427** and **431**, which absorb visible light (600 and 550 nm), recyclize within seconds even at −196 °C.[478,479]

The studies in Munich blended with those performed in other laboratories. The optically active oxirane **420** and its *cis* isomer served MacDonald and Crawford[470] as model. The racemization rates reflect those of electrocyclic ring opening, whereas *trans–cis* isomerization proceeds via rotation of the carbonyl ylides or by disrotatory ring opening. According to Paladini and Chuche,[480] *trans-* and *cis*-2-phenyl-3-propenyloxirane equilibrate at 240 °C and are irreversibly converted to *trans*-3-methyl-2-phenyl-2,3-dihydrofuran. Thus, rotation foils the recognition of the orbital-controlled mode of 1,5-electrocyclization.

Gary Griffin et al.[481] observed photofragmentation of aryloxiranes into carbonyl compounds and carbenes; in 1970, the colors observed on irradiation of *trans-* and *cis*-2,3-diphenyloxirane at −196 °C were attributed to the *trans–cis* isomeric carbonyl ylides. According to Lee,[482] irradiation in the presence of acetone as triplet sensitizer converted the same oxiranes in the presence of methyl acrylate to identical mixtures of *cis-* and *trans*-cycloadducts; rapid rotation of the triplet carbonyl ylides leads to stereorandomization. Griffin et al.[483] irradiated both *trans-* and *cis*-2,3-di-β-naphthyloxirane and obtained only the *cis*-2,5-dinaphthyl cycloadducts with *cis-* and *trans*-2-butene, whereas the stereointegrity of the alkene was preserved.

The cation radical derived from the carbonyl ylide by electron removal is likewise stereolabile. In 1978 Albini and Arnold[484] proposed a photoinduced electron transfer for sensitization by 1,4-dicyanonaphthalene (DCN). The cation radicals of *trans-* and *cis*-2,3-diphenyloxirane reached nearly rotational equilibrium of the open-chain cation radicals before back-donation of the electron from DCN$^{\cdot-}$ occurred. The retention of dipolarophile configuration lends credence to neutral carbonyl ylides entering 1,3-cycloadditions. The groups of Schaap[485] and Ohta[486] based an elegant formation of *cis*-3,5-diaryl-1,2,4-trioxolanes (ozonides) on the DCN-sensitized photooxygenations of both *trans-* and *cis*-diaryloxiranes.

After Theodor Curtius, at the age of 27, discovered diazoacetic ester (page 28), he asked his friend in the Munich laboratory, Eduard Buchner, his junior by 3 years (and subsequently Nobel laureate 1907), for cooperation. Their 1885 report[487] on a 1:2 product from ethyl diazoacetate and benzaldehyde (160 °C) was not a landmark. The structure was incorrect, and 25 years later Dieckmann,[488] likewise in Munich, recognized isomeric 1:2 products as the dioxolanes **434**. After preliminary tests we conjectured the intermediacy of a carbonyl ylide in 1963,[159] but embarked on a mechanistic study much later. Pedro de March, a guest from Barcelona, observed that dimethyl diazomalonate eliminated N_2 in benzaldehyde at a normal (not induced) rate; bis(methoxycarbonyl)carbene adds to benzaldehyde, and thus furnishes carbonyl ylide **437** that, in turn, combines with the second molecule of benzaldehyde. The yield of the diastereoisomeric 1,3-dioxolanes **436** was 56% in the thermolysis of diazomalonic ester at 125 °C and rose to 87% when N_2 elimination was catalyzed by Cu(I) triflate at 25 °C. Electrocyclization of **437** to give the oxirane **435** was competing with the cycloaddition (k_2/k_1 = 2.6 at 100 °C).[489]

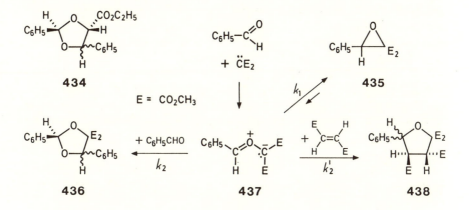

The double role of benzaldehyde was broken up by introducing dimethyl diazomalonate into a mixture of benzaldehyde and dimethyl fumarate at 125 °C. The two dipolarophiles competed for the carbonyl ylide **437** with k_2'/k_2 = 3.3 forming two diastereoisomers each of **436** and the tetrahydrofuran **438**.[490]

Beware of generalizations! On closer inspection and variation of the reactants by Franz Bronberger and Abhijit Mitra, the Buchner–Curtius reaction of diazoacetic ester and aromatic aldehydes led to a thicket of complications in rate and catalysis phenomena, as well as in product palette and steric course; the unpublished results[491,492] will not be sketched here.

Are any isolable carbonyl ylides known? Janulis and Arduengo[493] described the ylide **439**, which, however, does not behave like a 1,3-dipole because the partial structures holding positive and negative charge form an 80° angle. The crystalline mesoionic oxazolium-4-olate **440**, studied by Ibata[494] since 1974 and called "isomünchnone", is a better choice. It adds olefinic and acetylenic dipolarophiles as illustrated by **441**, the adduct of dimethyl fumarate.

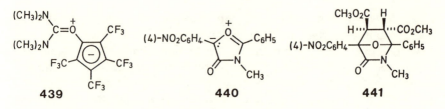

439   **440**   **441**

1,5-Electrocyclizations: An Exercise in Pigeonholing

Superlatives are dangerous. However, when commenting on the importance of electrocyclic reactions of the pentadienyl $\rightleftarrows$ cyclopentenyl
anion type in heterocyclic chemistry, a superlative is appropriate. Since
the mid-1960s I have collected examples in a fast-growing file. In 1980 I
proposed a consistent principle for ordering the somewhat chaotic a-
bundance and illustrated it by selected cases.[254]

The base-catalyzed cyclization of "hydrobenzamide" (**442**) to
amarine (**445**) was discovered by Laurent[495] in 1844. The Russian com-
poser and chemist Aleksandr Borodin[496] contributed to the structure of
445. In the 6π system of **443** the anionic charge is distributed over the
three C atoms and *migrates on cyclization*; the cyclopentenyl-type anion
444 holds the charge on the N atoms, the former centers 2 and 4. In a
beautiful study, Hunter et al.[497] established disrotatory cyclization
(>99.7%), **443** → **444**, as predicted by the Woodward–Hoffmann rules.

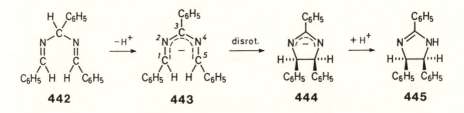

442   **443**   **444**   **445**

In 1970 my former coworker Hans Reimlinger[498] published a
paper on "1.5-dipolar cyclization", which constitutes the largest class;
two more reviews have appeared since.[499,500] In **446** a 1,3-dipole *a–b–c*
of the allyl type (**269**) bears a conjugating substituent *d=e* at the terminus. In the three octet structures the anionic charge is delocalized,
whereas the onium charge resides on *b*. Compared with the pentadienyl anion, the all-carbon analog, the neutral 2-CH is replaced by an
iminium or oxonium function; we called this *isoelectronic* exchange. Up
to four *isoionic* replacements—charge character preserved—are conceivable (Chart XIII). The formal charges of **446** disappear in **447** as a result
of the mentioned charge migration from *a,c,e* to *b,d*. Imino or oxo

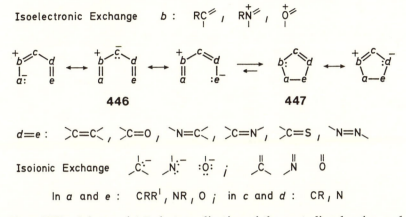

Chart XIII. Scheme of 1,5-electrocyclization of the pentadienyl anion and possibilities of isoelectronic and isoionic exchange.

groups as terminals possess lone pairs that modify the concert of motions in the electrocyclization.[254]

Subtle stereoelectronic problems occur in the cyclization of 1,3-dipoles of the propargyl—allenyl type (268) conjugated with a variety of $d=e$ groups. The chemistry of pentazoles, tetrazoles, and triazoles is rich in examples, such as 214 ⇄ 215, 216 ⇄ 217, 220 ⇄ 221, and 222 → 223 (pages 80–82). Phenylpentazole (215) is a cyclopentadienyl anion with one isoelectronic and four isoionic exchanges. The tautomerism of tetrazoles with imidazides, known since 1892,[501] is no more than one of a multitude of systems; a progress report 1965–1975 cited 134 references.[502]

Isoelectronic exchange in the 1 or 3 position allows charge-free systems to cyclize and give zwitterions. The N-benzoyl-N-methylphenylketene (313) cyclizes to münchnone 303, a pertinent case (page 103). Twofold isoelectronic exchange permits pentadienyl anion-type (6π) cyclizations to take place even in a *cation*.[254]

It is hard to beat the number of variables in this game of electrocyclization and impossible to exhaust it experimentally. In an 1842 letter to his friend Justus von Liebig, Friedrich Wöhler compared organic chemistry with "a tropical jungle, full of odd creatures, a vast thicket without exit or end." Complaints about the growing volume of chemistry and its increasing complexity have not subsided since. The jungle needs an infrastructure more than ever. "In the pentadienyl anion cyclization we are undoubtedly facing an archetype which permits the collective description of a colorful variety of reactions. The classification system introduced is suitable to pigeonhole hundreds of scattered examples from the literature."[254]

1,4-Dipolar Cycloadditions

A General Principle as a Contrast Program

The concept of 1,3-dipolar cycloaddition sprang from mechanistic considerations; mechanistic studies accompanied its systematic development. Concerted processes were designated—half in jest—as "no-mechanism reactions",[503] emphasizing the unavailability of *direct* evidence. The application of diagnostic criteria required *comparison standards* for two-step pathways via biradical or zwitterion. Previous studies from other sides did not attempt a synoptic view of a diversity of criteria. Searching for a contrast program, we encountered in 1,4-dipolar cycloadditions a general principle of a multistep process, yet we were unable to find good models for stereochemical and rate studies. The (2+2) cycloadditions of ketenes to olefins were our second choice. We ran into trouble by uncovering a novel concerted pathway (page 158). The third test case, the (2+2) cycloaddition of donor and acceptor olefins, allowed the desired collation (page 169). The borderline crossings (i.e., the two-step examples of 1,3-dipolar cycloadditions, page 117) were observed later (1985).

Concomitant with our endeavors, Jürgen Sauer, my former coworker, elaborated such a synopsis of criteria for Diels—Alder reactions.[382,504] Likewise concurrent were Paul Bartlett's[505,506] elegant and profound studies on the separation of concerted and biradical pathways in the cycloadditions of 1,3-dienes to fluorinated and chlorinated ethylenes.

Derived from scattered observations in the literature, the scheme of 1,4-dipolar cycloaddition that emerged in the early 1960s (Chart XIV)

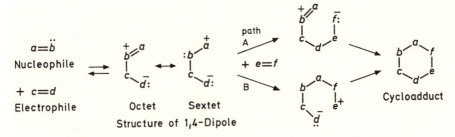

Chart XIV. Scheme of 1,4-dipolar cycloaddition; a–f may be carbon atoms or heteroatoms.

was distinctly different from the concerted pathway of the Diels–Alder reaction. The name was perhaps not an optimal choice because it suggested nonexistent mechanistic relations to 1,3-dipolar cycloadditions.

The combination of a nucleophilic and an electrophilic multiple-bond system generates the 1,4-dipole (i.e., a tetramethylene-type zwitterion that usually harbors one or more heteroatoms). The union with an electrophilic (path A in Chart XIV) or a nucleophilic bond system $e=f$ (path B) occurs in two steps and produces a six-membered ring. The frequently competing cyclization of the 1,4-dipole gives a four-membered ring. The zwitterions of type **492** from ketenes and enamines and **504** from TCNE and vinyl ether (pages 166, 171) are pertinent examples. Cyclic head–tail dimers of 1,4-dipoles have also been encountered.[507]

A Colorful Variety of Examples

Diels and Alder obtained labile 2:1 adducts from dimethyl acetylenedicarboxylate (DMAD) and pyridine[508] or isoquinoline;[509] a "transient species with two free valences", **448**, which adds across the C–N bond to form **449**, was conjectured. The initial combination of the heteroaromatic base with the electrophilic acetylenic ester appeared more probable, and Acheson and Plunkett[510] isolated a zwitterionic CO_2 adduct of **450**. I saw no reason why DMAD should play a double role in the formation of **449**; an interception of **450** by cycloaddition to a third multiple-bond system was conceivable.

Isoquinoline did not react with an excess of phenyl isocyanate or diethyl mesoxalate in the cold. Slow introduction of DMAD allowed the isolation of 1:1:1 products, which were structurally clarified by Klaus Herbig and Masanobu Morikawa: 46% of **451** and 70% of **452**, respectively. Erwin Brunn found dimethyl azodicarboxylate likewise a

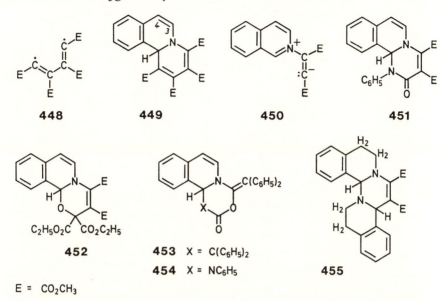

E = CO$_2$CH$_3$

suitable intercepting reagent for **450**.[511] We observed the formation of **453** (89%) from isoquinoline and diphenylketene at 20 °C and hypothesized a 1,4-dipole as an intermediate. In situ capturing by phenyl isocyanate to give **454** justified the assumption.[512]

Heteroaromatic bases forfeit their aromaticity in these cycloadditions. Azomethines reacted much faster. Depending on the mode of addition, 3,4-dihydroisoquinoline and DMAD furnished either the 1:2 product (3,4-dihydro derivative of **449**) or the 2:1 product **455**.[513] Similarly, the mixing mode of *N*-benzylidene-ethylamine and phenyl isocyanate determined whether the 1:2 or the 2:1 adduct (both with six-membered heteroring) were formed under conditions of kinetic control.[512]

The choice of the 1,4-dipolarophile is limited to electrophilic or nucleophilic multiple-bond systems. However, not in all examples was the electrophilic and nucleophilic component of the 1,4-dipole suitable as 1,4-dipolarophile (e.g., 3,4-dihydroisoquinoline with carbon disulfide or *N*-phenylmaleimide provided only 2:1 products, independent of the reactant ratio).[514,515]

In most of the cases studied, the 1,4-dipole is not isolable. An exception is the crystalline adduct **456** of *N*-benzylidene-methylamine and sulfur trioxide. Addition to the electron-rich ethyl vinyl ether and subsequent loss of ethanol afforded **457**,[507,516] as found by David S. Breslow, a guest from Hercules Inc., Wilmington.

Many casual observations in the literature can be interpreted as 1,4-dipolar cycloadditions, the first being 150 years old: V. Regnault[517]

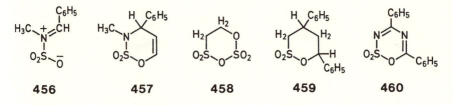

456   **457**   **458**   **459**   **460**

obtained "carbyl sulfate", **458**, from ethylene and sulfur trioxide. Interestingly, the reaction of styrene or benzonitrile with SO_3 gave rise to 2:1 adducts **459** and **460**.[518,519]

In 1,4-dipolar cycloadditions, it is still hard to predict whether a certain set of reactants will be successful or not. Sometimes modest structural alterations nullify the capacity. This capriciousness profoundly contrasts the nearly unlimited scope of Diels–Alder and 1,3-dipolar cycloaddition reactions. A fundamental mechanistic difference becomes evident. Electrophilic and nucleophilic termini of the 1,4-dipole are fixed, often separated by tetrahedral C functions; those of the 1,3-dipole are interchangeable. The difference in scope reflects the two-step pathway of the 1,4-dipolar cycloaddition and the concertedness of the 1,3-dipolar counterpart.

The unpredictability of the three-component reactions is cured by using cyclic mesoionic betaines as stable 1,4-dipoles. Such 1,4-additions to the malonyl heterocycles as pyrimidinium-, 1,3-oxazinium-, and 1,3-thiaziniumolates and their benzo derivatives were studied by Friedrichsen, Gotthardt, Kappe, and Potts in the 1980s. The 1985 review[349] by Ollis et al. lists a great number of cycloadditions of these and related 1,4-dipoles.

In the pyrimidinium-olate **461**, the negative and positive charge reside in the upper and lower half of the molecule. Gotthardt and Schenk[520] described the addition of ethyl cyanoformate to **461** giving **462**, which, in turn, undergoes cycloreversion at higher temperature. Interconversions like **461** → **463** are of synthetic value. The inventory of suitable dipolarophiles encompasses only electron-deficient and electron-rich multiple bonds corresponding to paths A and B of Chart XIV.

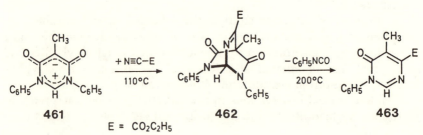

461   **462**   **463**

$E = CO_2C_2H_5$

Nearly 15 years was required for 1,3-dipolar cycloaddition to gain awareness of most synthetic chemists; the 1,4-dipolar counterpart still did not attain it. No help comes from textbooks; their authors are conservative and unduly neglect the entire field of cycloadditions.

(2+2) Cycloadditions of Ketenes

Cyclobutanones from Ketenes and Alkenes

In 1920 Staudinger and Suter[521] discovered the (2+2) cycloadditions of diphenylketene to styrene, vinyl ether, and cyclopentadiene. Perhaps the presumed limitation to "activated" alkenes was responsible for having neglected this versatile pathway to cyclobutanones over decades. We merely found the cycloadditions to *common alkenes* to be slower. Diphenylketene furnished 87% of **464**, R = C_4H_9, with 1-hexene at 100 °C in 8 h or 97% of **464**, R = CH_3, with propylene at 20 °C in 8 months;[522] butyl vinyl ether, however, afforded 99% of **464**, R = OC_4H_9, in 3 h at 20 °C.[523] NMR spectra and the conversion of **464** to 3-substituted 4,4-diphenylbutyric acids revealed the regiochemistry expected for a cycloaddition via zwitterion **465**. Capable graduate students were a blessing for the ketene project: Leander A. Feiler (Ph.D. 1967) was followed by Peter Otto (1970), Peter Koppitz (1974), and Herbert Mayr (1974).

The ring opening to **465** and its derivatives appears to dominate the chemistry of the cyclobutanones **464** (e.g., electrophilic catalysis produced the α,β-unsaturated ketones **466**). The formation of **467** from **464**, R = OC_2H_5, ($ZnCl_2$, benzene, 20 °C) is based on an intramolecular phenyl substitution of **465** and subsequent alcohol elimination.[523] Numerous conversions of 2,2-diphenylcyclobutanones into derivatives of naphthalene are known.

A puzzle is, "How many molecules of diphenylketene are bound by one molecule of water?" The surprising answer is three. The diphenylacetic anhydride combined with a third molecule in the pres-

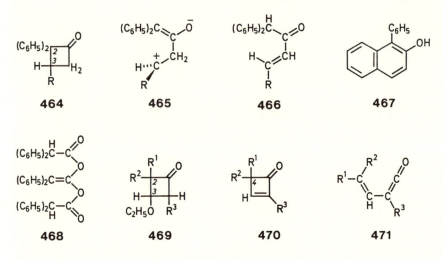

ence of a trace of triethylamine, affording the acylal **468** (1 month, 25 °C, 80%).[524]

A general synthesis of cyclobutenones **470** from adducts **469** simply consisted in the passage of an alumina column.[525] On warming, the cyclobutenones provided the highly reactive vinylketenes **471**.[526] The rate of this irreversible electrocyclic ring opening was similarly influenced by 2-substituents as the conversion of cyclobutenes to butadienes.[527]

Stereospecificity and Mechanism

The carbon of the carbonyl orbital of ketenes **472**, orthogonal to the projection plane, is the center of electrophilicity. A type-**465** zwitterion as intermediate in the addition to a nucleophilic alkene would profit from the heteroallyl anion resonance, and the conformation **473** possesses the lowest Coulomb potential. Stereoconvergence would be anticipated for ketene cycloadditions to *cis,trans* isomeric enol ethers via **473**.

In 1964 we reported on stereospecific additions of ketenes to *cis*- and *trans*-propenyl propyl ether.[528] No mutual admixture was observ-

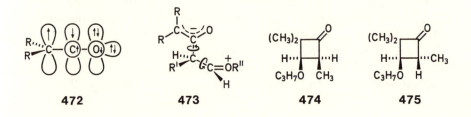

able in the *cis-* and *trans*-adducts of *diphenylketene*, but the NMR analytical limits, tested in artificial mixtures, amounted to several percent. The additions of *dimethylketene* to the same *cis,trans* pair also proceeded stereospecifically (>99%) without the "wrong" cyclobutanone becoming noticeable; here 0.8% of **474** and 1.2% of **475** were detectable by VPC in artificial mixtures.[529] According to Montaigne and Ghosez,[530] in situ cycloadditions of dichloroketene to *cis-* and *trans*-cyclooctene resulted in <1% of mutual admixture.

In their famous communication of 1965, Hoffmann and Woodward[311] listed ketene cycloadditions among those that are forbidden to be concerted by the orbital symmetry rules. When I informed R. B. Woodward of our findings, he proposed in 1967 a concerted $_\pi 2_s + _\pi 2_s + _\pi 2_s$ mechanism involving C=C and C=O of the ketene, sketched schematically in **476**. However, no symmetry element is retained in the process.[529]

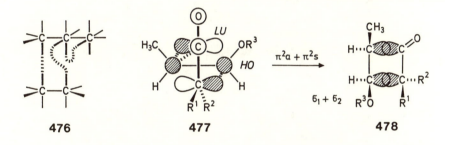

476   **477**   **478**

An intellectual adventure was the $_\pi 2_a + _\pi 2_s$ mechanism (C_2 symmetry), which was conceived for the concerted dimerization of two ethylenic bond systems and applied to the ketene cycloadditions by Woodward and Hoffmann[351] in 1969. Only by restricting the attention to the two bonds made and broken during the net cycloaddition, the C_2 axis is recognized in structure **477**. It illustrates the predominant contribution LU(ketene) + HO(ketenophile); the secondary orbital interaction with the low-lying LU(C=O), designated as the "spearhead", substantially lowers the transition-state energy. The rotations included in the conversion **477** → **478** cause painful van der Waals collisions of the substituents. Not a single example of the $_\pi 2_a + _\pi 2_s$ pathway has been secured by using normal olefins. The possession of only three substituents at the C=C bond of ketenes and the additional orbital interaction—a PMO calculation underlined its importance[531]—enable the ketene to play the *antarafacial* role in this odd mechanism. Because of the polarization of the carbonyl bond, ketenes resemble the bond system of the vinyl cation.

This may be an appropriate occasion to pay tribute to R. B. Woodward. His creativity and grasp of the essential, the wide scope of his interests, his profound knowledge of the literature, his apprehension, and his gift for combining seemingly unrelated phenomena are legendary 10 years after his premature death. With gratitude I remember many fruitful discussions that were not even interrupted while driving through heavy Cambridge traffic.

Woodward became a role model and created a new fashion, stereoselective synthesis of natural products; not all of his followers

Blue suit, blue tie; the now-legendary Robert B. Woodward (1917–1979) with son Eric in front of their home in Belmont, Massachusetts, in 1955.

were able to devise new reactions, tailor-made for the synthetic purpose. Woodward's strychnine synthesis (1963) showed the stroke of genius. In his B_{12} synthesis, he encountered peculiar steric courses of hexatriene cyclizations and recognized a novel phenomenon governing reactivity.[532] He was aware of the connection with the symmetry of the HOMO and asked Roald Hoffmann, a youthful Harvard Fellow, to cooperate in the study. This was the birth hour of the principle of conservation of orbital symmetry.

On personal contacts: Sports were not Bob Woodward's strong side, but I played tennis with him in 1955; neither of us was qualified for championship.

In 1971 the IUPAC Symposium on Cycloadditions in Munich was fortunate to win Bob Woodward as plenary lecturer. We, the organizers, followed a well-tried recipe by scheduling Woodward at the end of the agenda. That arrangement kept all of the participants in Munich until the end, and the program schedule was not derailed by an overlong lecture (3 h = 1 Woodward unit). Jürgen Sauer hired a brass quartet for a musical introduction to this climax of our meeting. The melodies were to be aired first outside as a signal to the people at the coffee table. Subsequently, the quartet would play on the gallery of our 700-seat lecture hall. We had miscalculated; the hall was packed and nobody left for the coffee break. All seemed to be too concerned about having a seat for the famous guest's lecture. Paul Bartlett gave a witty introduction. Woodward, after the quartet's splendid performance, decided to speak "in the future, never without music."

Woodward's style of lecturing was inimitable and his blackboard work impeccable. He always brought his own colored chalk. Posterity tends to disregard any bad habits of celebrities. To be quite frank, chain smoking during the lecture struck me as inappropriate.

When I presented the Arthur D. Little lectures at Northeastern University in Boston in 1974, I talked on ketene cycloadditions and octatetraene cyclizations, among other topics. I was moved by Woodward's taking the time to attend all three lectures and the banquet.

Structure–Rate Relationship: *cis* Effect

Structure **477** suggests a strong donor–acceptor interaction, but the transition state is far from being a full-fledged zwitterion. The cycloaddition rates of diphenylketene (benzonitrile, 40 °C) mirror the donor properties of the ketenophiles. 2,3-Dihydrofuran as a cyclic enol ether reacts 85,000 times faster than cyclopentene. In the sequence isobutenyl ethyl ether, N-isobutenylmorpholine, and N-isobutenylpyrrolidine, the rate constants increase 1:580:800,000.[533] The substituent effect is high, but still modest compared with that of electrophilic aromatic substitution. Bromination of anisole and N,N-dimethylaniline shows a rate ratio of $1:10^9$ despite the diminution of the substituent effect by the benzene ring.

The solvent dependence exceeds that of Diels–Alder reactions and 1,3-dipolar cycloadditions, but is much smaller than that observed for the (2+2) cycloadditions of tetracyanoethylene (page 176). The rate ratio in acetonitrile and cyclohexane amounts to 160 for butyl vinyl ether + diphenylketene and to 48 for 2,3-dihydropyran + diphenylketene; log k_2 values are fairly linear to E_T.

Diphenylketene combined with cis-2-butene to yield 96% of cycloadduct after 3 days at 90 °C, whereas the reaction with trans-2-butene was still incomplete after 3 months.[522] Competition of cis- and trans-propenyl propyl ether for diphenylketene revealed a k_{cis}/k_{trans} ratio of 170.[528,529] Effenberger et al.[534] found cis-1-butenyl ethyl ether even 295 times faster than the trans- isomer versus the same ketene. This unique superiority of cis-ketenophiles over the trans-isomers was mysterious at first. In Diels–Alder reactions and 1,3-dipolar cycloadditions, trans-disubstituted ethylenes exceed cis isomers in rate, the cis compounds suffering from steric hindrance of resonance.

The $_\pi 2_a + _\pi 2_s$ mechanism offers the clue to the singularly high cis–trans rate ratios. In the orthogonal arrangement of the orientation complex 477, one ketene substituent is placed between two vinylic ligands of the ketenophile. Diphenylketene combines with ethyl vinyl ether and ethyl cis-propenyl ether to form complexes 479 and 480, in which phenyl enters into van der Waals repulsion only with two vinylic hydrogens. In the addition to trans-propenyl ethyl ether, the ketene substituent must squeeze into the gap between an ethylenic substituent and vinyl-H, and the collision with the ethoxy group in 481 is the smaller evil. The retardation by a factor of 84 corresponds to $\Delta\Delta G^{\ddagger} =$ 2.7 kcal mol^{-1} (Chart XV). A second β-methyl in ethyl isobutenyl ether increases the donor activity, as attested by the ionization potentials. However, the anticipated rate increase is outweighed by the energy required to overcome the high interference of phenyl between methyl and hydrogen in 482; compared with 480, the rate is decreased 70,000-fold. The steric strain in the orientation arrangements 479–482 is exacerbated during the ensuing twist motions.[535]

The combination of alkylphenylketenes with cis- and trans-propenyl ethyl ether furnishes cyclobutanones with three stereocenters. A statistical analysis of the $\delta(^1H)$ for the ring protons of 59 cyclobutanones

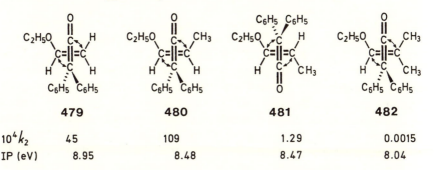

	479	**480**	**481**	**482**
$10^4 k_2$	45	109	1.29	0.0015
IP (eV)	8.95	8.48	8.47	8.04

Chart XV. *Rate constants for the cycloadditions of diphenylketene to ethyl vinyl ether and its β-methyl derivatives (benzonitrile, 40 °C, 10^4 k$_2$ in M^{-1} s^{-1}) and ionization potentials.*

provided substituent increments that allowed an unequivocal configurational assignment.[536] Each reactant pair can still afford two diastereoisomeric cyclobutanones when the configuration of the ketenophile is retained. In Chart XVI the measured rate constants are split according to product composition for the pairs of orientation complexes, **483/484** and **485/486**. The ratios reflect the steric requirements of the ketene's inward substituents: CH_3 < C_2H_5 < C_6H_5 < $CH(CH_3)_2$, $C(CH_3)_3$. The *cis–trans* rate ratios k_B/k_D pertain to the reactions via phenyl inward orientation complexes **484** and **486**; k_B/k_D amounts to 164,

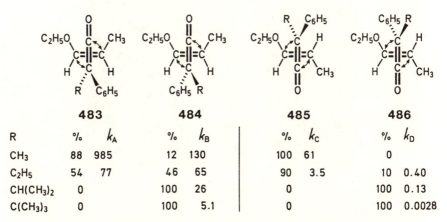

R	**483**		**484**		**485**		**486**	
	%	k_A	%	k_B	%	k_C	%	k_D
CH$_3$	88	985	12	130	100	61	0	
C$_2$H$_5$	54	77	46	65	90	3.5	10	0.40
CH(CH$_3$)$_2$	0		100	26	0		100	0.13
C(CH$_3$)$_3$	0		100	5.1	0		100	0.0028

Chart XVI. *Cycloadditions of alkylphenylketenes to ethyl cis- and trans-propenyl ether in benzonitrile at 40 °C; rate constants (10^6 k$_2$ M^{-1} s^{-1}) partitioned within the pairs according to percent product composition.*

200, and 183 for R = ethyl, isopropyl, and *tert*-butyl, respectively; it is virtually independent of the size of R, which juts outward. The pathways via **483** and **485** with R inward are no longer used when R is isopropyl and *tert*-butyl because the steric congestion in the transition states is too great.[537] Herbert Mayr was a highly efficient co-worker in this demanding preparative and kinetic project.

Characteristically, diphenylketene adds stereospecifically to the *cis* double bond of *cis,trans*-2,4-hexadiene;[538] *trans*- and *cis*-1-methylbuta-diene accepted diphenylketene exclusively at the unsubstituted double

A prolific associate, Herbert Mayr, received his Ph.D. in 1974. After a postdoctoral fellowship in George A. Olah's laboratory he studied cycloadditions of allenyl cations for his habilitation at Erlangen (1980). Mayr became a professor at the Medical University at Lübeck in 1984 and moved to Darmstadt in 1991. The alkylation of alkenes by carbenium ions and a new probe for distinguishing one-step and two-step cycloadditions are his present research fields.

bond to give 3-*trans*-propenyl- or 3-*cis*-propenyl-2,2-diphenylcyclobuta-none.[539]

Cyclobutanone Formation from 1,3-Dienes

Ketenes do not add as dienophiles to the 1,4 positions of 1,3-dienes, but rather yield cyclobutanones exclusively. Why should an open-chain intermediate **487** show such a predilection for the four-membered compared with the six-membered ring? In the framework of the $_\pi 2_a + _\pi 2_s$ pathway, the regiochemistry shown by **488** is the result of partial allyl cation and olate anion character in the transition state. The secondary orbital interaction in **477** likewise explains the following rate preference: diphenylketene reacts 30,000 times faster with cyclopentadiene than with cyclopentene.[539]

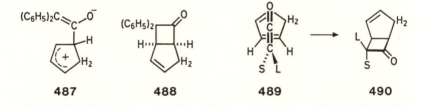

| 487 | 488 | 489 | 490 |

In 1970 no less than four research groups studied cycloadditions of ketenes with a large and a small substituent (L and S) to cyclopentadiene. The predominance of the adduct **490** with the larger substituent in *endo* position was regarded as evidence for the $_\pi 2_a + _\pi 2_s$ pathway.[540] The orientation complex **489** with S between the vinylic hydrogens guides the process to the bicyclic ketone **490**.

A renaissance of ketene chemistry commenced in the late 1960s. An excellent 1977 review by Ghosez and O'Donnell[540] also covers the (2+2) cycloadditions of ketene iminium salts, which share mechanistic features with the additions of ketenes.

Ketene cycloadditions are no "orchid chemistry", no exotic curiosity. The cycloadduct of dichloroketene to cyclopentadiene allowed a rational synthesis of tropolone, as described by Stevens et al.[541] in 1965. Ghosez and his group used the cycloaddition of ketenes and nucleophilic cleavage of cyclobutanones to effect the stereospecific addition of two carbon chains to the double bond of cyclopentadiene.[542] In 1981 this vicinal dialkylation based on the addition of (methoxycarbonyl)chloroketene opened an economic route to prostaglandins.[543]

Enamines as Ketenophiles: Two Mechanistic Pathways

Dimethylketene reacts with N-isobutenyldialkylamines to give cyclobu-tanones and 2:1 products; the latter were clarified by Hasek et al.[544] as δ-lactones of type **493**.

We found the cyclobutanone **491** as a 1:1 adduct stable to di-methylketene, and the dependence of the product ratio, **491:493**, on the reactant ratio suggested **492** as a common intermediate.[545] It came as a surprise that rate measurements and competition experiments by Peter Otto demonstrated a *dichotomy of pathways* for dimethylketene + N-iso-butenylpyrrolidine.[546]

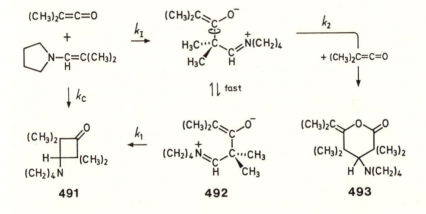

Quantitative yields of **491** + **493** were obtained with 6–7 equiv of dimethylketene in increasing molar concentration (acetonitrile, 25 °C). A plot of the ratio of mono- and bisadduct, **491:493**, versus the recipro-cal dimethylketene concentration gave a straight line that did not go through the origin as expected for ring closure (k_1) and reaction with the second mole of dimethylketene (k_2) competing for the intermediate. The intercept (0.78 = 44:56) indicated that only 56% of the enamine was converted to an intermediate that chooses between the paths with k_1 and k_2; the remaining 44% of the enamine reaches the cyclobutanone **491** on the concerted pathway with k_C, thus avoiding the mentioned competition.[546]

In the reaction system discussed, the cyclobutanone **491** ori-ginates from a concerted and a two-step pathway, the latter via an iminium zwitterion **492**. It is meaningful to postulate such an inter-mediate for the reaction with the most nucleophilic ketenophile. The formation of δ-lactone **493** is portrayed as a 1,4-dipolar cycloaddition (page 151) of **492** to dimethylketene. Steady-state treatment of this kinetic system leads to eq 3 (i.e., a linear dependence of the cyclobutanone–δ-lactone ratio on the reciprocal dimethylketene concentration D with k_C/k_I as intercept).[547] In equation 3, k_C is the rate constant for the concerted process and k_I is the rate constant for zwitterion formation.

$$491:493 = \frac{k_C}{k_I} + \frac{k_1}{k_2}\left[1 + \frac{k_C}{k_I}\right]\frac{1}{D} \qquad (3)$$

This reaction scheme is more than science fiction. Similar series of experiments were carried out in six additional solvents. The percentage of the pathway via zwitterion **492** ranged from 8% in cyclohexane to 56% in acetonitrile. The photometric rate constants were divided into k_C and k_I. When the solvent was changed from cyclohexane to acetonitrile, the rate of the concerted process (k_C) was increased 37-fold, whereas a 540-fold growth of k_I denotes the greater charge separation in the transition state of zwitterion formation.[546]

Should the zwitterionic pathway not also be opened by making the ketene more electrophilic instead of increasing the nucleophilicity of the ketenophile? According to England and Krespan,[548] bis(trifluoromethyl)ketene and ethyl vinyl ether in hexane at 0 °C furnished the oxetane **494**, which at 50 °C rearranges to cyclobutanone **495**. A zwitterion of type **473** must be involved; without rotation it can do no more than close the oxetane ring.

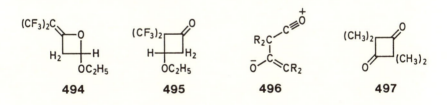

Dimerization of Ketenes and Formation of β-Lactams

The dimerization of ketenes to give methylene-β-lactones or 1,3-cyclobutanediones belongs to the classic inventory of ketene chemistry. The oxonium olate zwitterion **496** is a plausible intermediate for the interconversion of the two types of dimers, but does it occur in the dimerization? Dimethylketene forms solely **497**; we found the rate constant of dimerization in acetonitrile at 35 °C 30 times higher than in CCl_4.

This value suggests an unsymmetrical transition state. The moderate size of the solvent effect, and the Eyring parameters ($\Delta H^{\ddagger} = 10.8$ kcal mol^{-1}, and $\Delta S^{\ddagger} = -42$ eu in benzonitrile) are better in harmony with a $_\pi 2_a + _\pi 2_s$ pathway than with the formation of a zwitterion 496.[549]

In 1907 Staudinger[550] discovered β-lactams by reacting ketenes with azomethines. He likewise observed 2:1 adducts with dimethylke-tene.[551] These adducts were recognized 60 years later by Martin et al.[552] as 2-methylenehexahydro-1,3-oxazine-6-ones of type 500. When we introduced diphenylketene slowly into N-benzylidene-methylamine in acetonitrile, NMR analysis indicated 82% of lactam 498 and 6% of the bisadduct 500. The inverse procedure, slow addition of the azomethine to diphenylketene, yielded 19% of 498 and 81% of 500; the β-lactam 498 is resistant to diphenylketene. The 1,4-dipole 499 is an attractive common intermediate that either cyclizes or interacts with a second molecule of diphenylketene.[553] Quantitative experiments by Pavol Kristian, a guest from Kosice (Slovakia), made competition between a concerted and a two-step pathway in the formation of 498 very probable.

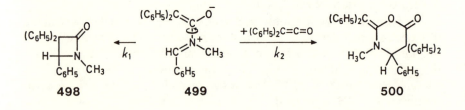

(2+2) Cycloadditions of Donor and Acceptor Olefins

The Tetramethylene Problem

All experimental evidence for concertedness in cycloadditions is indirect and results from comparison with authentic two-step processes (page 151). We were on the lookout for nonconcerted cycloadditions amenable to mechanistic scrutiny. However, our study of cyclobutane formation from donor and acceptor olefins developed beyond a mere "contrast program" to 1,3-dipolar cycloadditions; it included five Ph.D. theses stretched over 1969–1986[555].

The dimerization of ethylene is an equilibrium reaction with ΔH = −19 kcal mol^{-1} and ΔS = −46 eu; the concentration of cyclobutane amounts to 99.8 mol% at 25 °C, but shrinks to a sheer nought of 0.01 mol% at 500 °C. High temperature is required to overcome an activation barrier of 44 kcal mol^{-1} for the dimerization, but now the dissociation is favored in the equilibrium.

That phenomenon raises the mechanistic question. Because $_\pi 2_s$ + $_\pi 2_s$ is forbidden to be a thermal concerted process, a two-step reaction via a 1,4-biradical or 1,4-zwitterion remains. Ethylenic substituents that stabilize carbon radicals or cation + anion should lower the energy level of the intermediate and make the (2+2) cycloaddition feasible.

Tetramethylenes of the *biradical* or *zwitterion* type appear in the MO treatment as extremes on a continuous scale.[556] Nevertheless, it is practical to denote the prevailing character by the term 1,4-biradical or zwitterion. Ab initio calculations on the tetramethylene parent disclosed a preference for the *anti* over the *gauche* conformation, **501** and **502**, on a rather flat surface.[557] Through-bond coupling between the p

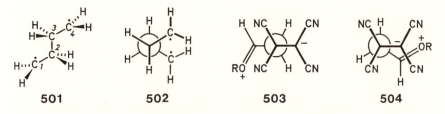

501   **502**   **503**   **504**

orbitals at the termini and the σ orbital of the 2,3-bond impedes rotation
about the 1,2 and 3,4-bonds.[558] The height of this barrier depends to a
disturbing extent on the method of calculation; values between 2.8 and
45 kcal mol^{-1} have been recorded.[559,560] Thus, the experimenter does
not get much help here from theoretical chemistry.

Ted Cairns and his group at Du Pont beautifully developed poly-
cyanocarbon chemistry. Besides tetracyanoethylene (TCNE), the 1,1-
dicyano-2,2-bis(trifluoromethyl)ethylene (**505**) was prepared.[561] Furthermore, Proskow, Simmons, and Cairns[562] made the 1,2-isomers **506** and
507 accessible by an ingenious method; the last step was the pyrolysis of
a chlorosulfite in sulfur vapor at 460 °C.

Visit at Central Research Laboratory, E. I. du Pont de Nemours and Company,
Wilmington, 1957. From left: John D. Roberts (consultant), Blaine C. McKusick,
me, Ted L. Cairns, David C. England, and Howard E. Simmons (later director of
research). The laboratory was nicknamed the American Industrial University.

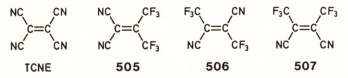

TCNE	**505**	**506**	**507**

In 1952 Williams et al.,[563] working in the same laboratory, described the formation of a cyclobutane from TCNE and ethyl vinyl ether at 25 °C. They suggested a pathway through a zwitterionic intermediate.

These four tetra-acceptor-substituted ethylenes served as models in elaborating mechanistic criteria for the (2+2) cycloaddition via 1,4-zwitterion. The cyclization of the zwitterion requires the *cisoid* or *gauche* conformation **504**; but is this the initial species? The conformational equilibrium of the 1,4-zwitterion from TCNE and vinyl ether, **503** ⇌ **504**, will be influenced by Coulombic attraction that should favor *gauche* over *anti*.

Donor Rotation and Dissociation of Zwitterion

The cycloaddition of TCNE to ethyl *cis*-propenyl ether proceeds quantitatively. The *cis* relation of methyl and ethoxy was retained in the main product **510**, X = O. The percentage of *trans*-adduct **511**, X = O, formed by cyclization after rotation about the former donor bond (**508** → **509**, X = O) increased from 5% in benzene to 15% in acetonitrile (Chart XVII).

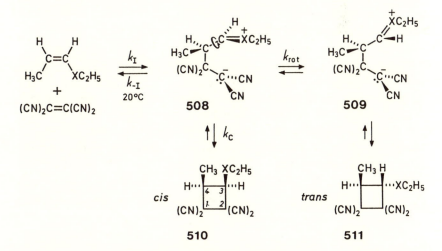

Chart XVII. *Kinetics and stereochemistry for the (2+2) cycloadditions of TCNE to ethyl* cis-*propenyl ether and sulfide (X = O,S).*

Conversely, on starting with ethyl *trans*-propenyl ether, *cis*-adduct **510**, X = O, constitutes the result of stereochemical leakage, its share rising from 4% in benzene to 23% in acetonitrile. Coulombic attraction between the charge centers of **508** and **509** will be diminished in media of high dielectric constant; a greater mobility is the consequence.[564] Thus, rotation competes with cyclization, but rotational equilibrium is not attained.

In the addition of TCNE to *cis,trans* isomeric ethyl propenyl sulfides, some inversion was likewise observed.[565] In highly polar solvents the cyclization of the zwitterion is reversible. In 2 M $LiClO_4$ in ether (resembling a salt melt[566]) *cis-trans* equilibria are rapidly attained; the *cis* share **510** amounts to 18% for X = O and 7% for X = S.[565]

After the reaction of TCNE with an excess of *cis*- or *trans*-1-butenyl ethyl ether in acetonitrile, the unconsumed enol ether was partially *cis,trans* isomerized. Numerical evaluation revealed that *cis*- and *trans*-zwitterion undergo ring closure (k_C) and dissociation (k_{-I}) at nearly the same rate (Chart XVII).[567] The ratio is >1 for *cis*- and *trans*-propenyl ethyl sulfides, and cyclization is generally preferred to dissociation in less polar solvents.

Gerd Steiner's splendid work (Ph.D. 1972) was more than a prelude to the polyphonous tetramethylene theme. The musical collation is fitting: Steiner, a chemist at the BASF AG, composes symphonies in his spare time, emboldened by his pianist wife. Reinhard Schug (Ph.D. 1976), the second player, successfully concentrated on the interception reactions.

Structure–Rate Relationship

The charge-transfer absorption enables us to measure the cycloaddition rates photometrically. Equation 4 shows that the experimental second-order rate constant k_2 is a composite of k_I for zwitterion formation and a partition coefficient of the high-energy intermediate. When the structure of the enol ether or thioenol ether is varied, the step with the late transition state (k_I) is probably more affected than those with early transition states (k_C and k_{-I}) in Chart XVIII.

$$k_2 = \frac{k_I k_C}{k_C + k_{-I}} \tag{4}$$

The (2+2) cycloadditions of TCNE are rather selective. 1-Alkenes are inert, vinyl ethers are suitable, and their structure–rate relationship is in harmony with a late transition state resembling the carboxonium intermediate of type **504**.[568] α-Alkylvinyl ethers (2-ethoxypropene, 1-ethoxycyclopentene) already react too fast for conventional measurement. 1,1-Disubstituted butadienes or *trans*-fixed 1,3-dienes are amenable to cyclobutane formation with TCNE.[569,570] Enamines interact rapidly with TCNE, but produce substitution products or tars.

Surprisingly, the k_2 values of thioenol ethers are similar in magnitude to enol ethers.[571] No doubt the conjugation energy of a vinyl ether exceeds that of a vinyl sulfide. However, when it comes to stabilizing a full carbocationic charge, the sulfur function is at least as effective as oxygen, a phenomenon understood by theoreticians.[572,573]

The rate ratios of TCNE additions to *cis*- and *trans*-propenyl ethers varied only from 1.2 to 0.76 and those of the sulfides from 0.26 to 0.13.[568,571] The contrast to the cycloadditions of ketenes (pages 161–165) is marked and underlines a fundamental mechanistic disparity.

$$H_2C=CH-Do \;+\; R_2C=C(CN)_2 \;\underset{k_{-I}}{\overset{k_I}{\rightleftharpoons}}\; \underset{R_2C-\bar{C}(CN)_2}{\overset{H_2C-\overset{+}{C}=Do}{|}} \;\overset{k_c}{\longrightarrow}\; \text{(cyclobutane)} \quad i \quad k_2 \approx k_I \text{ in nonpolar media}$$

Rate Constants $10^3 k_2$ at 25°C

	$H_2C=CH-OC_4H_9$	$H_2C=CH-SCH_3$
BTF (R = CF$_3$), benzene	2 700	4 900
TCNE (R = CN), benzene	1.1	0.60
Ratio BTF/TCNE	*2 500*	*8 200*

	$H_2C=CH-OC_2H_5$	$H_3C-\overset{H}{C}=\overset{H}{C}-OC_2H_5$	$H_3C-\overset{H}{\underset{H}{C}}=C-OC_2H_5$
BTF, benzene	2 900	6.5	4.9
TCNE, ethyl acetate	19	40	53

Reaction Enthalpy ΔH_r (kcal mol^{-1}) in CH_2Cl_2 at 25°C

	$H_2C=CH-OC_4H_9$	$H_2C=CH-SC_6H_5$	
BTF (R = CF$_3$)	− 27.1	− 26.0	$\Delta\Delta H \approx 5$
TCNE (R = CN)	− 22.0	− 21.1	kcal mol^{-1}

Chart XVIII. Rate constants and reaction enthalpies for (2+2) cycloadditions of TCNE and BTF at 25 °C.

Cyano stabilizes a carbanion to a higher extent than CF$_3$. Therefore, (2+2) cycloadditions of 1,1-dicyano-2,2-bis(trifluoromethyl)ethylene (BTF, **505**) to vinyl ethers take place exclusively via the better zwitterion affording cyclobutanes with alkoxy and cyano functions in adjacent positions. Middleton's report[561] about BTF reacting with methyl vinyl ether at −78 °C electrified us. According to measurements by Reinhard Brückner[574], (Chart XVIII), BTF adds to butyl vinyl ether 2500 times faster than TCNE, and versus methyl vinyl sulfide the ratio even reaches 8200. Why should BTF be a *superelectrophile*? When TCNE + donor olefin furnish the zwitterion, two cyano groups are decoupled from conjugation; in contrast, CF_3 does not show a resonance effect in BTF. Can this phenomenon account for the startling reduction of $\Delta G^{\ddagger}$ by 4.6 or 5.3 kcal mol^{-1}, respectively?

We measured the reaction enthalpies for the (2+2) cycloadditions of BTF and TCNE to two donor olefins (Chart XVIII). The cycloadditions of BTF turned out to be *more exothermic* than those of TCNE by roughly 5 kcal mol^{-1}.[574] Two cyano groups lose conjugation in the first

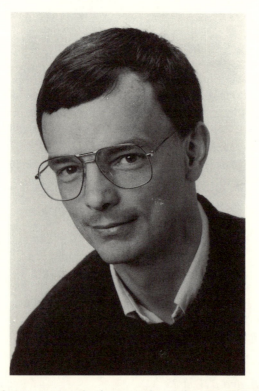

Reinhard Brückner, a very productive co-worker, delivered a Ph.D. thesis (University of Munich, 1984) of nearly 700 pages. He habilitated at Marburg in 1989 and moved as associate professor to Würzburg in 1991. Senkrechtstarter (fast risers) being scarce in Germany, Brückner obtained a full professorship in Göttingen in 1992. His main interest is stereocontrolled synthesis.

case, four in the latter. However, further factors must participate in generating an effect of that dimension. Fluorine as substituent is known to change C–C bond energies, σ as well as π.[575]

The exceptionally high rate of BTF (**505**) in its (2+2) cycloadditions is all the more astonishing because steric encumbrance at the CF_3-bearing C atom should retard zwitterion formation. This steric hindrance becomes obvious on introducing β-methyl into ethyl vinyl ether; whereas k_2 of TCNE is slightly increased for ethyl *cis*- or *trans*-propenyl ether, k_2 of BTF was diminished 450- and 600-fold (Chart XVIII).[574]

Two β-methyl groups in ethyl isobutenyl ether (**512**) still allowed a smooth (2+2) cycloaddition of TCNE; that of BTF was thwarted, however, by massive hindrance. Instead, *hydride abstraction* from the allylic position furnished the ion pair **514**, which collapsed to afford **515**. The same product **515** of formal allylic substitution arose from **505** + ethyl methallyl ether (**513**), supporting the ion pair intermediate **514**.[576]

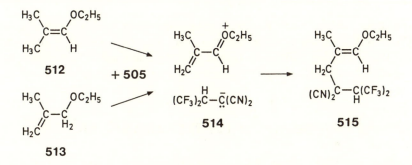

Kinetic comparison of (4+2) and (2+2) cycloadditions of polycyanoethylenes is enlightening (Chart XIX). Stepwise introduction of further cyano groups into acrylonitrile up to TCNE increases the rate constant of Diels–Alder reactions with cyclopentadiene 41 millionfold, the k_2 with 9,10-dimethylanthracene even 15 billionfold as described by Sauer et al.[577] The log k_2 values are linearly related to the differences of the ionization potentials of diene and the electron affinity of the dienophile,[578] in accordance with an *early transition structure* (TS) of the concerted process.

In contrast, the (2+2) cycloadditions of 1,1-dicyano-, tricyano-, and tetracyanoethylene to isobutenyl methyl ether decrease slightly. The late TS resembles the zwitterion, which is stabilized by two cyano groups in all three cases.[579] Acrylonitrile and fumaronitrile are unreactive because the zwitterion is insufficiently stabilized by only one CN group. In terms of absolute rate constants, the (4+2) addition of TCNE to dimethylanthracene is 3 billion times faster than its (2+2) addition to

	(2+2) Benzene, 25°C	(4+2) Dioxane, 20°C	
Acrylonitrile	0	1.04	0.89
Fumaronitrile	0	81	139
1,1-Dicyanoethylene	31.6	45,500	127,000
Tricyanoethylene	2.4	480,000	5,900,000
Tetracyanoethylene	2.0	43,000,000	13,000,000,000

Chart XIX. (2+2) and (4+2) cycloadditions of polycyanoethylenes; rate constants 10^5 k_2 $M^{-1}s^{-1}$).

isobutenyl ether, thus demonstrating "the magic of concertedness" (Chart XIX). To be concise: the rate data present a criterion for early and late transition states (the title of our communication); concerted and stepwise cycloadditions are reasonable interpretations.

Solvent Dependence of (2+2) Cycloadditions

The rates of Diels–Alder reactions and 1,3-dipolar cycloadditions show a small or negligible response to solvent polarity.[298e,504] However, TCNE cycloadditions exceed even Menschutkin reactions in the influence of solvent polarity on rate. We measured k_2 values in a range of solvents, from the nonpolar CCl_4 to the polar acetonitrile at 25 °C. The following rate ratios reflect the increasing charge separation in the activation process, in harmony with the late TS of the endothermic formation of a zwitterion.[571,580]

Electron-Rich C=C	k_{CH_3CN}/k_{CCl_4} for TCNE
3,4-Dihydro-2H-pyrane	17,000
Ethyl isobutenyl ether	4,900
Butyl vinyl ether	1,700
Ethyl cis-propenyl sulfide	17,100
Ethyl isobutenyl sulfide	2,900

Thus, increasing polarity of the medium diminishes $\Delta G^{\ddagger}$ by up to 5.8 kcal mol^{-1}. In contrast with TCNE, BTF (**505**) has a dipole moment of 1.74 D,[574] that is, cycloadditions of BTF start at a higher level of polarity. The lower polarity difference between ground and transition state compared with TCNE is responsible for a smaller solvent dependence.

Electron-Rich C=C	k_{CH_3CN}/k_{CCl_4} *for BTF*
Ethyl *cis*-propenyl ether	770
Butyl vinyl ether	133
Phenyl vinyl sulfide	540

Solvation is an area where chemistry becomes too difficult for chemists, and *empirical parameters* of solvent polarity provided their rescue. The rates of the (2+2) cycloadditions were measured in 8–15 solvents and the log k_2 gave fairly linear relations with the Reichardt parameter E_T.[581] The reaction TCNE + ethyl isobutenyl ether was measured in methanol and ethanol as well, yet these log k_2 values grossly deviated from the straight line. A four-parameter equation provided an excellent fit, including the alcohols. Only two were important: polarization, $(\epsilon - 1)/(2\epsilon + 1)$, i.e., the Coulombic term accounted for 72% of the solvent effect and 22% was nucleophilic solvating power, the latter with a negative sign originating from complexation with TCNE.[547] Hermann Graf (Ph.D. 1980) did much of the kinetic work; he was well versed in computer programs and juggled the solvent-effect data in multiparameter correlations.

$$\ln k_2 = \ln k_0 - \frac{1}{k_B T} \frac{\epsilon - 1}{2\epsilon + 1} \left[\frac{\mu_A^2}{r_A^3} + \frac{\mu_B^2}{r_B^3} - \frac{\mu_{\ddagger}^2}{r_{\ddagger}^3} \right] \tag{5}$$

Equation 5 was developed by Laidler and Eyring,[582] based on a crude electrostatic model by Kirkwood. This equation describes the rate constant for the reaction of spherical dipolar molecules A and B as a function of the dielectric constant of the medium; μ and r are dipole moments and radii, respectively, of substrates and TS. Previous applications of eq 5 were hampered by neglect of specific solvation forces. Because of the preeminence of the polarization term in our (2+2) cycloadditions, fair correlations of log k_2 with the function of the dielectric constant ϵ were obtained. Dipole moments of 10–14 D were calculated by means of eq 5 for the transition states of TCNE cycloadditions to four enol ethers.[580] This μ range is roughly two-thirds of the estimates for the *gauche* zwitterions 504; μ values of *trans* zwitterions of type 503 should be much larger (31–34 D). Further arguments for the initial formation of *gauche* zwitterions will be recounted.

Interception of Tetramethylene-Type Zwitterions

Trapping is an elegant way of establishing the occurrence and nature of an intermediate. Protic bases HB should be capable of intercepting the 1,4-zwitterions, but they also combine with the acceptor olefin. In 1971 I learned from Howard Simmons at a Gordon Conference that TCNE is resistant to cold pure alcohols. When Gerd Steiner combined ethyl *cis*-propenyl ether and TCNE in methanol, the zwitterion 516 was indeed captured by the solvent and converted to the mixed acetal 519. The methanolysis of cyclobutane 518 likewise afforded acetal 519, but this reaction at 25 °C was 47 times slower than the formation of 519 from the *cis*-propenyl ether and TCNE.[583] Thus, it is an intermediate of the cycloaddition that is trapped in the fast reaction with methanol.

Under conditions of kinetic control in methanol at 0 °C, zwitterion 516 cyclized to the extent of 11%, and 86% of 519 was found; in addition, 3% of a diastereoisomeric acetal, 520, occurred. The choice of different alkoxy groups in *cis*-propenyl ether and the intercepting

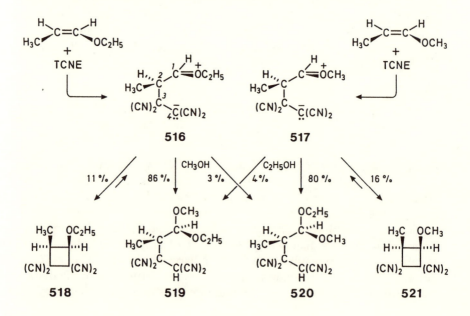

alcohol disclosed a remarkable stereoselectivity of the trapping reaction. An X-ray analysis of a related acetal by I. Karle et al.[584] secured the structures of **519** and **520**. The slow S_N1-type methanolysis of cyclobutane **518** necessarily passes the *cisoid* or *gauche* conformation **516**; with 99% of **519** and 1% of **520**, its stereospecificity was slightly higher.[585,586]

Provided the zwitterion **516** has a lifetime sufficient for rotation about C-1–C-2, the acetals **519** and **520** should be formed in comparable amounts. The selectivity in the formation of **519** and **520** is regarded as evidence for both the fast interception and the *gauche* conformation of the carboxonium zwitterions **516** and **517**, respectively.

If the *gauche* form **516** of the tetramethylene zwitterion is favored over the *trans* conformation by Coulombic attraction, is it feasible to pull the charge centers apart in a strong electric field? K. Wisseroth at the BASF AG, Ludwigshafen, had the equipment for studying reactions in high electric fields. My co-worker Hermann Graf went to Ludwigshafen and reacted TCNE with ethyl *cis*- and *trans*-propenyl ether in a field of up to 6×10^7 V m^{-1}. A lower stereospecificity in the formation of cycloadducts **510** and **511** would indicate the participation of *trans* zwitterions of type **503**. However, the results were inconclusive;[587] not all dreams come true.

A casual observation by Reinhard Schug led to the interception of the zwitterionic intermediates by 1,4-dipolar cycloaddition (page 151). The reaction of TCNE with ethyl vinyl ether in acetonitrile rendered 4% of the tetrahydropyridine **524** and 96% of the cyclobutane **522** under conditions of kinetic control (10 min, 20 °C). After 3 weeks at 20 °C, the solution contained 99% of **524**. The cyclobutane was gradually siphoned off through a small equilibrium concentration of the open-chain zwitterion **523** that adds to acetonitrile, furnishing the more stable six-membered ring of **524**. Analogously, the reactants combined in acetone to give **522** and **525** in 94:6 ratio; 1 week later a quantitative yield of **525** signaled thermodynamic control.[588]

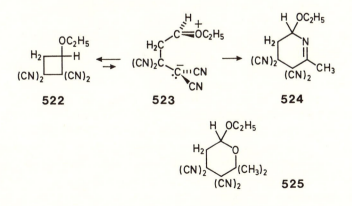

The substituents determine where a specific tetramethylene species has to be located on the continuous scale from biradical to zwitterion (page 169). The observations described here are consistent with zwitterionic reactivity and need no recourse to the diradical. Spontaneous polymerization occurs upon mixing certain donor and acceptor olefins. According to Henry K. Hall,[589,590] the quickly formed tetramethylene is the initiator. The formation of an *alternating copolymer* indicates diradical reactivity of the tetramethylene, whereas the zwitterion gives rise to ionic propagation with the cationic or anionic *homopolymer* as product. This valuable tool signals the nature of the initiator, and the polymerization chain serves as amplifier.

A "borderline crossing" was observed when the electron-rich thiocarbonyl ylide **332** reacted with tetra-acceptor-substituted ethylenes: (3+2) cycloaddition via a 1,5-zwitterionic intermediate (page 117). Yet in the case of dimethyl fumarate, a stereospecificity of >99.97% in the cycloaddition of **332** suggested a concerted course. The Hall criterion likewise confirmed concertedness for the thiolane formation from **332** with acrylonitrile or methyl acrylate: Cycloaddition was not accompanied by anionic polymerization.[591] One or two acceptor substituents in the dipolarophile are simply not sufficient for breaking the "spell of concertedness" of 1,3-dipolar cycloadditions.

Acceptor Rotation

In their (2+2) cycloadditions to butyl vinyl ether, *trans*- and *cis*-1,2-bis(trifluoromethyl)ethylene-1,2-dicarbonitrile (**506** and **507**) are slower than TCNE by factors of 17 and 10, respectively.[592] Two cyano groups stabilize a carbanion better than $CN + CF_3$. The ethylene derivatives **506** and **507** served as probes for acceptor rotation in (2+2) cycloadditions. The 1966 report of Proskow, Simmons, and Cairns[562] on stereospecific additions of **506** and **507** to ethyl vinyl ether, carefully studied by Gonzalo Urrutia (Ph.D. 1985), needs revision. The ^{19}F NMR spectroscopy of 1966 is not comparable with high-resolution technique in the 1980s. The additions to methyl vinyl ether may be singled out.[593].

The four diastereoisomers of **526** (three stereocenters) were obtained from *trans* and *cis* acceptor olefin in different ratios under conditions of kinetic control (Chart XX)[593]. The four cycloadducts were separated and structurally clarified by X-ray[594] and ^{19}F NMR data. Equilibration required drastic conditions (ethereal 2 M $LiClO_4$, 4 months at 70 °C). The equilibrium concentrations reflect the repulsion of voluminous *cis–vic* substituents: 80:20 for *trans* and *cis* located CF_3 groups; the diastereoisomer with 2-CF_3 and 3-OCH_3 in *cis* position is less favored within each pair.

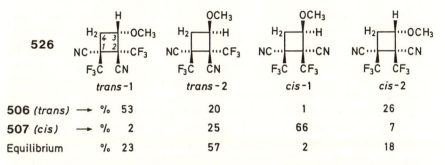

		trans-1	trans-2	cis-1	cis-2
506 *(trans)*	→ %	53	20	1	26
507 *(cis)*	→ %	2	25	66	7
Equilibrium	%	23	57	2	18

Chart XX. (2+2) *Cycloadditions of acceptor olefins* **506** *and* **507** *with methyl vinyl ether in dichloromethane at 25 °C, percent composition of cycloadduct (yield 93–94%); trans and cis refers to the vicinal CF$_3$ groups.*

The *trans*- and *cis*-configurations of **506** and **507** are retained in 73% of their cycloadducts **526** (Chart XX). The major retention products, 53% of *trans*-1 from **506** and 66% of *cis*-1 from **507** are the thermodynamically *less favored* isomers with respect to the 2,3-substituents.[593]

Zwitterions **527** and **528**, the precursors of cycloadducts *trans*-1 and *trans*-2, are illustrated by their conformations with smallest distance of ionic charges; more of the carbanionic charge is passed on to CN than to CF$_3$, lifting the charge center toward CN. The sum of Coulombic and conformational strain favors 1S,2Re,3Re in **527** over 1S,2Re,3Si in **528**, *Re* and *Si* being the descriptors of two-dimensional chirality.[595] Similarly, *cis*-1 comes from **507** via the favored zwitterion 1S,2Si,3Si. Thus, on cyclization *with retention*, the better zwitterions give rise to the less advantageous cycloadducts.[593]

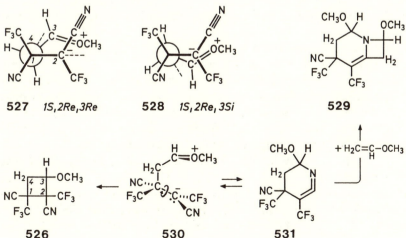

The major *inversion* products, 26% *cis*-2 from **506** and 25% *trans*-2 from **507**,[593] originate from the same zwitterions as *trans*-1 and *cis*-1, respectively, by cyclization after rotation about the acceptor bond. Initially formed *gauche zwitterions* (in contrast to *anti* of type **503**) are the basis of our interpretation, and we conclude that acceptor rotation is faster than donor rotation in our zwitterions.

Sometimes hard work is rewarded by fringe benefits. When **506** was reacted with methyl vinyl ether used in large excess as solvent, the yield of cyclobutanes **526** dropped to 41%, and Gonzalo Urrutia isolated 53% of 1:2 adducts **529**.[593] The structures of four diastereoisomers of **529** were elucidated, two of them by X-ray.[594] Because **526** is stable to methyl vinyl ether, a six-membered cyclic ketene imine **531** is required as an intermediate that adds vinyl ether at the CN double bond. The highly strained **531** reverts to zwitterion **530** if the concentration of the donor olefin is insufficient. The ketene imine **531** originates from a 90° stop of the acceptor rotation in **530**. The ratios of the four diastereoisomers of **529** buttress the conclusions on the preference of *ReRe* and *SiSi* zwitterions drawn from the cyclobutane structures.

The role of charge-transfer complexes and of single-electron transfer in our (2+2) cycloadditions is still unclear. A mechanistic investigation resembles a shaft that leads into a mine. Equipped with modern physical and intellectual tools, the chemist reaches a greater depth today than 50 years ago, but the range is still limited.

Further Reactions of Tetra-Acceptor-Substituted Ethylenes

TCNE and acceptor ethylenes **505**–**507** are potent dienophiles. However, the product from 2-methoxyfuran and **506** was neither a (2+2) nor a (4+2) cycloadduct. To Urrutia's and my surprise, a cinnamic ester with a tetra-acceptor-substituted cyclopropane in the *cis*-β-position emerged; a 77:23 mixture of **533** and **535** was isolated.[593,596] The 1,6-zwitterion **532** is a likely intermediate; before or after rotation in the acceptor part it apparently experiences an intramolecular nucleophilic substitution with the carboxylic ester as leaving group. The substitution product **534** resulted when the interaction took place in the presence of pyridine, the zwitterion **532** now being stabilized by 1,3 prototropy.

In the (2+2) cycloadditions of **506** and **507** to methyl vinyl ether, the stereochemical leakage amounted to 27%. The (4+2) cycloadditions of tetra-acceptor-substituted ethylenes to 1,3-dienes could conceivably pass zwitterions harboring allyl cations. However, the Diels–Alder reactions to 2,3-dimethylbutadiene and cyclopentadiene revealed high

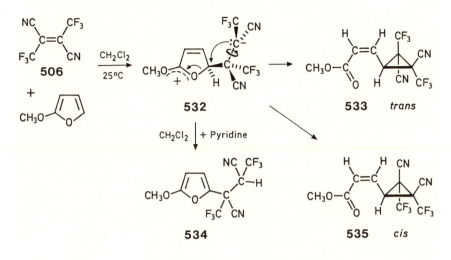

stereospecificities (Chart XXI).[597] The indicated lower limits were established by artificial mixtures (e.g., addition of 200 ppm of **537** to the pure *trans* adduct **536** gave rise to a visible GC peak). The GC peak of **537** was absent in the product from 2,3-dimethylbutadiene and **506**; thus retention of configuration must exceed 99.98%.

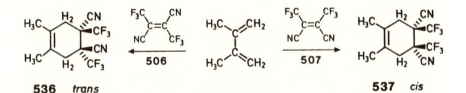

Stereospecificity (GC): **506** *(trans)* > 99.98% ; **507** *(cis)* 99.88%

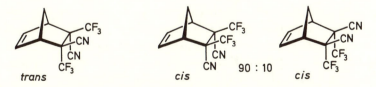

Stereospecificity (GC): **506** *(trans)* > 99.97% ; **507** *(cis)* 99.82%

*Chart XXI. Diels–Alder reactions of 2,3-dimethylbutadiene and cyclopentadiene with acceptor olefins **506** and **507** in dichloromethane at 25 °C; product analysis by capillary gas chromatography.*

Considering the sensitivity of the *cis* isomer **507**, purified by triple preparative GC, we are not certain in ascribing 0.12% and 0.18% of the *trans* adducts, respectively, to a minor nonstereospecific pathway. We assume that the concerted (4+2) cycloaddition, allowed by orbital control, exceeds the rate of the two-step pathway by a large margin. Sauer et al.[598] reported 99.98% stereospecificity for the Diels–Alder reactions of cyclopentadiene with fumaronitrile and maleonitrile.

Like TCNE, BTF (**505**) is capable of electrophilic aromatic substitution. *N*-Methylpyrrole was converted to **539** via zwitterion **538**; furan also underwent α-substitution, furnishing a side chain with terminal nitrile groups. When 2,5-dimethylfuran was treated with **505**, the propensity of the latter for hydride shift led to an attack on the methyl group, affording the ion pair **540**. The recombination product **541** contains terminal CF_3 groups.[599] Hence, the two substitution mechanisms gave rise to different end groups.

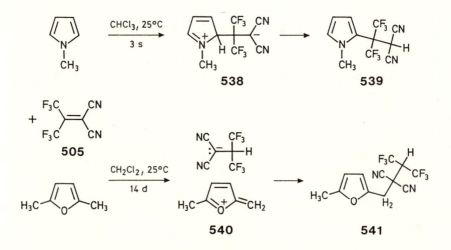

BTF (**505**) as a superelectrophile forces *common alkenes* into *ene reactions* at room temperature. Reinhard Brückner and I were baffled to find ene products with either terminal CF_3 or CN groups.[575] When we wrote the communication several years later,[600] our blindfold was lifted. There are two discrete ene pathways, the first controlled by C–C bonding (transition structure **542**), and the second determined by transfer of allylic hydrogen (TS **543**). In pictograms **542** and **543** the alkene is reduced to the backbone in which the bond reorganization takes place. Partial charges in the TS result from the different extent of C–C and C–H bond formation (···· weaker than ||||). The superiority of CN over CF_3 in stabilizing the partial negative charge determines the orientation.

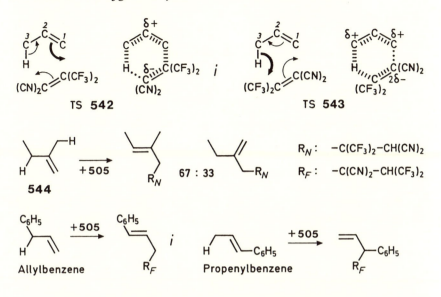

Propylene and BTF reacted 93:7[561] via TS **542** and **543**. Introduction of phenyl or alkyl into position 2 of propylene should stabilize the partial positive charge in **542**, whereas the same substituents in 1 or 3 position favor the quasi allyl cation in **543**. The orientations predicted on this basis were observed in all of the dozen alkenes studied.

Isobutene or 2-methyl-1-butene (**544**) afforded only products with terminal CN; in **544** the C–H of methyl and ethyl compete for incorporation into the cyclic process. Allylbenzene and propenylbenzene are 3- and 1-phenylpropylene; the pathway via TS **543** was used exclusively. If the allylic hydride were fully removed in the first step, the ion pair would give rise to identical products. The obligatory double bond shift underlines *concertedness* and *kinetic control*.

Glimpses and Reflections: More on My Life and Thoughts

Childhood and Adolescence

I was born in 1920 in Gerolstein, the son of Dr. Edmund Huisgen and Maria Huisgen, born Flink. I grew up in that little town in the Eifel, the highlands between the Rhine and Mosel Rivers. As a child I was shy and rather an outsider. This characteristic may have been a result of my bronchial asthma, which demanded long absences from school. The illness, combined with my innate disposition, fostered individualism. I shared my classmates' enthusiasm for soccer only half-heartedly, and I was a poor team player. All my life I have remained somewhat of a loner, not engaging in cooperative research as freely as I could have.

From the age of 10 I attended the classical secondary school, where I learned Latin, classical Greek, and French (no English). I was more attracted by mathematics and biology than by language studies.

My father had a strong interest in science, and I was fascinated to hear about biological evolution, measurement of the speed of light, and heavy water (a brand-new discovery at that time). At the age of 13 I started to collect fossils. Gerolstein is located in rich Devonian strata, and a local geologist helped me to identify my finds of brachiopods, crinoids, and trilobites.

My first contact with chemistry came through a small experimental kit, a Christmas present I received at the age of 14. Like many youngsters, I pulverized potassium chlorate with red phosphorus and managed to scorch my hair. The way I devoured popular science books led my parents to joke about their "little professor". Luckily, the uninspiring high school chemistry instruction was limited to 1 year, not enough to squelch my enthusiasm.

187

Carnival in Gerolstein in 1926. My father, Dr. Edmund Huisgen (1888–1939), with my brother, Klaus (left), age 8, and me, age 5, both in disguise. Obviously I did not feel relaxed in the role of Little Red Riding Hood.

As I approached the end of high school my father, a surgeon, allowed me to watch him operate. He advised me to study medicine. However, my admiration for my father's extraordinary skills made me doubly conscious of my own average gifts in that respect, and I decided against it.

As often happens, the turning points in my life were decided by chance. After my *Abitur* (final high school exam, 1938) I plowed through the 6 months of *Reichsarbeitsdienst* (labor service) obligatory in Hitler's Germany; however, I contracted pneumonia at the end. Because of my convalescence, the usual enlistment order was temporarily suspended. Bad luck, good luck! In the spring of 1939 I was given the chance to go to the University of Bonn.

At age 4 I dreamed of becoming a sailor, but other interests soon took over.

I was 12 years old when Hitler seized power in 1933. The world-wide Depression was a catastrophe for Germany, which was still paying reparations to the Allies because of World War I. Disastrous unemployment rates fueled an increasingly aggressive desperation, which in turn was probably partly responsible for the country's eagerness to line up behind Hitler's promise of economic change. All institutions were rapidly streamlined by the new rulers, and political indoctrination became a major educational goal. Youngsters were often torn between the political influence of their school and the opposing spirit they experienced at home. I remember my father once speaking of Joseph Goebbels, Hitler's minister of propaganda, as the devil incarnate; I contradicted him at the time.

People in free countries find it hard to understand how individu-

Maria Huisgen (1890–1978) with her two sons, Klaus (left) and Rolf (right), about 1932. The dolomite rocks at Gerolstein were a climber's paradise.

als may live with a dictatorial regime, seemingly acquiescent, in order to maintain some freedom of thought. In my own case, the personal confidence I had in my father increasingly outweighed the pressure of school and the media. Needless to say, I never joined the NSDAP (Hitler's party), but I am sorry to add that neither did I engage in active resistance. When Hans and Sophie Scholl and their *Weisse Rose* (White Rose) group risked—and lost—their lives in open political opposition in 1943, their act appeared to us, the passive resisters, as foolhardy rather than heroic. Only much later did I begin to fully value their self-sacrifice as a beacon of light during a dark age. Now I view the "strategy of avoidance" in a different light. I consider it one of the principal reasons that more than half of the human race still bears the yoke of oppressive regimes.

More on My University Years

I immersed myself in the intellectual adventure of learning. In the first place, I profited from attending lectures; they set a good pace and reduced my awe of the texts. Furthermore, the external pressure of the

constant uncertainty about when I would be enlisted acted as an ac-
celerator. An additional time-compressing factor was the trimester system at Munich, one of the four German universities that remained open
at the beginning of World War II in September 1939; three terms were
crowded into the year instead of two.

During my first three terms mathematics and chemistry ranked
equally high in my interest. With the growing demands both fields
made on my time, a decision was inevitable. As a high school boy I had
dreamed of studying many of the natural sciences and humanities. My
dream popped like a balloon as I understood the need to narrow the
scope of my curiosity.

Nevertheless, I felt a strong desire at least to survey the major
areas within organic chemistry. That goal was not easy to reach 50
years ago and is even more difficult today. A high degree of specialization is the tribute we pay to the dimensions of an expanding discipline.
Unfortunately, specialization is a little like looking at the sky from the
bottom of a well—the deeper the well, the narrower the outlook. Many
scientists resignedly look for "research niches" that will offer life-long
shelter.

However, chances that prolonged research will be truly worthwhile in the confines of such a shelter are poor. I remember from early
ventures into paleontology that a certain brachiopod survived in an
ecological niche since the late Cambrian period (several hundred million
years), but it was a cul de sac in evolution.

My father died in 1939, and I eked out a living from a fraction of
his life insurance. After the first exams I obtained a tuition waiver, and
Heinrich Wieland helped me by providing free chemicals and glassware;
later a small teaching assistantship followed.

After five trimesters I passed the exam required for the diploma
and started research under Wieland's guidance. He offered me a problem involving biological oxidation. I did not feel up to it yet and chose
an alkaloid topic because I wanted to lay a solid foundation in organic
chemistry. Despite my interest in the life sciences, organic chemistry
never loosened its grip on me after that.

Hitler's initial order permitted research only if it was likely to
help win the war within a single year. This order was later revised, and
many scientists were called back from the front. In the summer of 1942
I was conscripted to military training for 4 months. Wieland succeeded
in retrieving me, although my research was unrelated to the war efforts.
My affliction with juvenile asthma contributed to a temporary dis-
charge.

I was one of Wieland's last doctoral students; perhaps I did not
profit as much from the master's experience and guidance as did those
of earlier years. Administering an institute during World War II was a

highly demanding job, particularly for someone of Wieland's fragile health. Nonetheless, time permitting, he was strongly involved in counseling his co-workers at the bench.

I received my Ph.D. in 1943. Wieland was feared as a strict examiner but fortunately he was in a good mood, puffing away at his cigar. He even offered one to his examinee, who judiciously declined. Afterward my lab mates were amused about the story because, in idiomatic German, "to hand out a cigar" means "to reprimand".

A Note on My Thesis Work

Wieland and his group had investigated vomicine, a minor companion of the alkaloid strychnine in *Strychnos nux vomica*, since 1928 in nearly 30 publications. My manipulation of the functional groups, especially the seven-membered unsaturated ether ring and the basic nitrogen, led to a multitude of new derivatives and stereoisomers but did not provide insight into the puzzle of the condensed rings.

In 1936 Leopold Horner, one of my predecessors in the field (later a professor at Mainz and highly reputed for his organophosphorus and o-quinone studies), had degraded vomicidine (lactam carbonyl reduced to CH_2) to vomipyrine, which was regarded as **546**, a derivative of the aromatic 3*H*-pyrrolo[3,2-*f*]quinoline, on the basis of nearly identical UV spectra.[601] My related degradation of vomicine afforded oxyvomipyrine with the corresponding 2-quinolone system. My synthetic efforts were misled by the mentioned spectroscopic similarity.

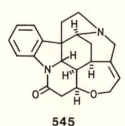

545

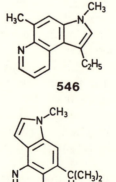

546

547

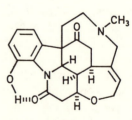

548

In fresh approaches after 1945, Prelog, Robinson, and Woodward succeeded in making the last correction leading to the strychnine formula **545**,[3] later buttressed by two X-ray analyses. Structure **548** would correspond to vomicine. The conclusion that vomipyrine was derived from the 7H-pyrrolo[2,3-h]quinoline skeleton was unavoidable; the central benzene ring contains the two nitrogen functions in 1,3 positions instead of 1,4 as originally assumed. Interestingly, the precursor subjected to the palladium dehydrogenation did not yet contain the pyrrolidine ring. While we were still working on the nitration of *p*-cymene after the installation of our stopgap laboratory at Weilheim, Robinson and Stephen[602] published an article on the synthesis of vomipyrine (**547**) via dinitro-*p*-cymene. In 1963 R. B. Woodward's breath-taking total synthesis of strychnine[603] crowned the endeavors that had begun as early as 1818 with the isolation of the alkaloid.

I am far from holding adverse conditions—maybe a crowded air raid shelter does not provide optimal conditions for creative thinking—responsible for my lackluster findings on the strychnine problem. At the age of 22 I was not experienced and mature enough to crack one of the hardest nuts of alkaloid chemistry. Children sometimes retain an aversion to books that are beyond their intellectual capacity when they first tackle them. I wonder whether similar reasons prevented my return to natural products after forays into other fields.

The Uncushioned Postwar Years

Because the Institute buildings in Munich had been destroyed in air raids, I was put in charge of installing a wooden shack as a training laboratory for organic chemistry in Weilheim, Upper Bavaria (page 5). Although municipal offices were helpful, the obstacles encountered were substantial. An anecdote may illustrate the living conditions of a chemist during the first postwar years. How did we acquire the wooden benches, the plumbing, and the electrical installation for the new laboratory? Money was available, but the Reichsmark was not worth much. Thus, we supplemented our cash payments with booze; a saved barrel of pure ethanol worked wonders. I presented the "statements of account" to H. Wieland.

Many stories blossomed about chemists' survival strategies after the war. Obviously, similar ideas sprang up everywhere. The sweetener dulcin (*p*-ethoxyphenylurea) was easy to prepare and was traded for potatoes and eggs. Soap-making required more expenditure; the moderate quality of the product was marketable only because the soap obtained with ration cards was even worse. Ethanol for chemical purposes was denatured by petroleum ether; much imagination was devoted to the removal of the petroleum ether, and the methods were closely guarded.

A sudden economic recovery set in when financial reform brought the Deutsche Mark in 1948. The German *Wirtschaftswunder* ("economic miracle") was based on the people's dogged determination to rebuild the country, materially and politically. Has a stable political, social, and economic system been achieved? In scientific terms, it is metastable at best. *Wunder* strikes me as an adequate description of Germany's situation in 1990: It had the world's highest wages, shortest work hours, longest vacations, and lavish unemployment compensation; nevertheless, it was the world's foremost export country. Unfortunately, the *Wunder* was not permanent; see "Bonn's Boom Goes Bust" (*Time*, May 24, 1993).

On Role Models

Guest speakers were rare during the war but, after a pause of several years, seminars were resumed on a regular basis. I enjoyed listening to the lectures of scientists with whom I was quite familiar through the literature. On occasion my respect grew into admiration when closer contact revealed not only a high intellectual capacity, but also a likable personality.

The list of my early role models is headed by *Heinrich Wieland* (page 4). Along with passing on high standards in research, Wieland set an example of scientific and human integrity. He had a warm-hearted personality; his self-discipline and unbending sense of duty made a major impression on the younger people around him. Wieland courageously resisted the tide of political intolerance and opportunism. In particular, he made great efforts to protect scientists and students from racial persecution,[604] an effort that often succeeded because of his prestige as a Nobel Laureate. A lively portrait of Wieland, based on recollections and original documents, was recently drawn by Bernhard Witkop, one of his former students.[605]

In the early postwar years Wieland presented the introductory course in the anatomy theater, one of the few big lecture halls that could be used without an umbrella on rainy days. I was responsible for the experimental demonstrations. In the summer of 1947 Wieland, who had a chronic injury to his right hand, could not write on the blackboard. I did the writing instead, and the formula often appeared on the chalkboard just before Wieland mentioned the compound. He wagged a warning finger at me—with a smile—and in the subsequent year asked me to present the course.

Heinrich Wieland (1877–1957) was a master in the art of crystallization, which requires patience and rubbing with a glass rod. Or did the ashes of his cigar play a catalytic role? (Photograph in 1943 by Bernhard Witkop.)

As my interest in reaction mechanisms evolved in 1947–1950, I adopted additional role models. *Hans Meerwein* wrote as few as 70 papers, 15 past the age of 70. A classic paper with van Emster in 1922[108] presented the kinetics of equilibration of bornyl chloride, isobornyl chloride, and camphene hydrochloride; the high solvent influence and the catalysis by protic and Lewis acids were conspicuous. Meerwein inferred an ionization and located the skeletal change in the carbocations. The echo was minimal in 1922, and a related paper was rejected as "too speculative". Alkoxo acids, trialkyloxonium salts, and $Al(OR)_3$-catalyzed Meerwein–Ponndorf reduction were further accomplishments. Recognition came after the war, with an abundance of honors.

When I first met Hans Meerwein in Marburg in 1950, the 71-year-old radiated enthusiasm. He told me of his new ventures in diazo

Hans Meerwein (1879–1965), Marburg, was an excellent storyteller; here he is chatting with Trudl Huisgen after receiving an honorary degree from the Faculty of Science, University of Munich, 1958.

chemistry and approved of my own attempts in this area. On the side, I learned from him that Paul D. Bartlett of Harvard University had sent him all-important CARE packages without even knowing him personally. The gesture was generous indeed.

In 1951, the Marburg faculty nominated me as Meerwein's successor. However, the minister of education in Hesse did not comply with my reasonable request for a modest increase in the institute's deplorable budget. In 1958 the faculty of science of the University of Munich, following my suggestion, granted the degree of Dr. honoris causa to Hans Meerwein; he lectured on carboxonium salts. The 79-year-old still mesmerized his audience.

Rudolf Criegee (1902–1975) of Karlsruhe appealed to me by the clarity of his mechanistic deductions as well as by his integrity and unassuming personality. The concise style of his reports often veiled his endeavors and experimental mastery. Criegee's elucidation of the ozonolysis pathway (pages 96–97) is textbook knowledge. Further feats were the glycol cleavage by lead tetraacetate, the conversion of alkenes into *cis*-diols by osmium tetroxide, cyclobutadiene chemistry, and new fundamental insight into the autoxidation of alkenes. An excellent and highly disciplined lecturer, Criegee preferred chalk and blackboard to a slide show. When the University of Munich celebrated its 500th

Rudolf Criegee wrote his publications as "success stories" transmitting the impression of an author blessed with life-long good luck; his scientific opus is a treasury. His former co-workers, many of them in academic life, and his friends remember him with reverence.

anniversary in 1972, our faculty gave Rudolf Criegee an honorary degree. Only a few years later I had the sad obligation of writing an obituary for him.[606]

From 1957 to 1975, as one of the five editors of *Chemische Berichte*, I cooperated with Rudolf Criegee, who had the final word on acceptance or rejection of papers. Usually authors received the editors' comments anonymously, but in tough cases Criegee composed his letter with particular diplomacy and actually signed it. As is well known, editorial activity is hardly ever immediately rewarding, and it has the additional "charm" of being extremely time-consuming. Nonetheless, I do not regret my time investment on that score. Unquestionably, the quality of a journal is proportional to the competence and dedication of its editors.

My acquaintance with *Paul D. Bartlett*, my senior by 13 years, dates back to the early 1950s; it developed into a durable friendship that included both families. It appears that our antennae were similarly tuned; as a consequence, our discussions were stimulating and enriching throughout. Habitually, I stopped at Harvard—later at Texas Christian University—on my trips to the United States. Paul Bartlett spent long periods in Munich (6 months in 1957 and 3 months in 1977), much to my delight and the benefit of my group.

The mere enumeration of Bartlett's contributions—bridgehead reactivity, hydride transfer from alkanes to carbocations, addition and chain transfer in polymerization, and photooxidation—does not reveal his elegance in designing crucial experiments and choosing models. Bartlett's studies on cycloadditions ran parallel to ours, a situation that gave rise to delightful discourse.

The diploma of Dr. rer. nat. h.c. is written in Latin. The Faculty of Chemistry and Pharmacy, University of Munich, honored Paul D. Bartlett, one of the fathers of physical organic chemistry, in 1977. He has educated generations of chemists. Some American physical organic chemists even boasted that they were not Bartlett's students, in order to show their independence.

Paul Bartlett is a nature lover. His home in Weston, Massachusetts, was located in a forest—part of his property—where we strolled and occasionally cut down rotten trees. We went skiing together in the Italian Dolomites, hiked up Mount Lafayette in Vermont, and biked along Trinity River at Fort Worth, Texas. In recent years I visited Paul in Watertown near Boston and in 1990, after Lou Bartlett's decease, in Lexington, Massachusetts. Again we walked; Paul was well above 80 and bravely fighting physical frailty. After Paul's retirement I was asked by the faculty of Texas Christian University in 1987 whether I was interested in taking over his Robert A. Welch Chair; I felt too deeply rooted in Munich to accept.

Chemical bonds link both atoms and the people studying molecules. My first meeting with *John D. Roberts* in 1955 was followed by many more on both sides of the ocean. I was fond of his constructive criticism, which pushed his interlocutor toward disclosure of all supporting arguments—or the lack of such. Roberts's readiness to concentrate in depth on his discussion partner's problem, no less than on his own, served me as an example. I have always considered it an intellectual game to listen actively, to anticipate arguments when taking in reports on new reactions or mechanistic pathways.

Roberts's clarification of reaction mechanisms by ^{14}C labeling and his inroads into the application of NMR spectroscopy and MO theory

Jack Roberts at sunrise on Zabriskie Point. On a joint excursion in the Spring of 1959 I found my guide less informed on Death Valley than on benzyne chemistry; his only previous visit had been at the age of 4.

were important stimuli. A 12-hour lecture course on molecular orbital calculations, which Roberts presented in Munich in 1962, was attended by both graduate students and staff.

Role models need not be elders. *George A. Olah*, my junior by 7 years, first visited in Munich around 1954. Two years later he fled from Budapest during the Hungarian Revolution and settled in the United States. My respect for his creative energy, willpower, and productivity has grown into a personal tie. A friendship among George and Judy Olah, my wife Trudl, and me was solidified by visits in Cleveland, Ohio, Los Angeles, and Munich. In 1973 Olah gave a course on *Elektrophile Reaktionen* (electrophilic reactions) in the Munich Institute. George is polyglot, switching from one language to another with mind-boggling speed.

Olah's onium chemistry is now an integral part of the college chemistry program. In particular, his ingenious idea of generating and stabilizing carbocations in superacids proved to be a major breakthrough. Moreover, the distinction between carbenium and carbonium ions with coordination numbers of 3 and 5, respectively, became meaningful once Olah accumulated evidence for the hypervalent species.

With George A. Olah (left) discussing the role of doubly protonated species in organic chemistry in Beverly Hills in April 1992. (Photograph by Trudl Huisgen.)

One can only marvel at Olah's output of nearly 1000 research papers and 11 books. Certainly, the reputation of being a very prolific writer may be a dubious one. To remove all doubt in Olah's case, I refer to the *Science Citation Index*, which measures a scientist's impact reliably. Olah is, in fact, among the world's most frequently quoted organic chemists.

What makes us praise novel achievements as imaginative or highly original? Facts and opinion from the literature are available to all of us, and our ideas are always influenced by them. However, those opinions can build up to massive prejudices. The judicious weighing of criteria and arguments is very personal, as is the process of connecting seemingly unrelated phenomena and extrapolating from them into the unknown. Scientific imagination is not so much wild fantasy that is completely detached from the existing body of experience as it is *absence of prejudice* about what can be done and what cannot.

In this respect, my friend *Heinz Ross* was an important model for me. He offered a paradigm of how life can be mastered under adverse conditions. While studying chemistry, he had set up a small laboratory on the side to prepare chemicals to order. At the age of 30 he lost his eyesight in a peroxide explosion. Gifted with incredible willpower, Ross nonetheless completed his Ph.D. program and managed his small enterprise specializing in fine chemicals. He closed down his laboratory at age 65 and is now attending classes in history and the history of art at the University of Munich, not even shying away from exams. With an excellent memory at his disposal, he has built a "tapotheque" (collection of recording tapes) and devised his own principles of retrieval; he is thus in a position to carefully prepare his seminar talks. My friend feels no vocation to be a teacher, but he has certainly taught me a lesson in steadfastness and optimism.

On Family Life

My fiancée, Trudl Schneiderhan, passed her Ph.D. exam in chemistry early in 1945; one of the required oral examinations took place in an air raid shelter. Having begun her studies at the University of Freiburg, Trudl opted for Munich in September 1939 when most German universities were closed. It was there that Trudl and I became acquainted in the laboratory in the early 1940s.

Together we watched Munich crumble under the air raids that started in 1943, a rather inhospitable setting for our romance. Jointly we removed rubble and broken glass from the laboratory benches and reglazed the windows. However, after direct hits by incendiary and demolition bombs in the fall of 1944, the Institute buildings at Sophien Street were abandoned. A couple of months later Trudl and I were bombed out of living quarters as well; our lodgings in different parts of the city both vanished during one night's inferno. Subsequently, Trudl and I moved to Weilheim, 50 km south of Munich, where we were married in July 1945, a few months after the end of the war. We thus followed good chemical tradition; "chemical marriage bonds" were frequent in Germany at the time and still are.

As for our backgrounds, Trudl came from Swabia, I from the Rhineland; in other words, we came from opposing camps. Germany is one of many countries plagued by long-standing north–south antagonisms. The imaginary boundary formed by the River Main separates two regions that have not always been on the best of terms, to say the least. Nowadays, however, the controversy mainly provides material for jokes, the best of them originating in our family, of course.

How much of a private life can a dedicated, if not to say singleminded, scientist expect? How much if both partners are professionals? I was lucky and got far more than expected. My wife Trudl gave up her professional career in favor of mine. As a consequence, she devoted much more time than I did to the raising of our two daughters, Birge (1946) and Helga (1949).

In 1949 our family of four moved from two small rooms in Weilheim to an actual house in Tübingen. The university had put up eight wooden houses for professors, drab little places by today's standards, but royal in our eyes. This arrangement made child-raising a good deal easier because the house came with its own little garden, where we could deposit our offspring when they became too obstreperous. There was also some free amusement built in. The homes were identical, so one had to keep count while climbing up Mörike Street. On occasion our neighbor, the paleontologist Schindewolf, would inadvertently enter our hall and each time be alarmed at the sight of a baby buggy.

In 1952, after some months of searching, we succeeded in finding a convenient and spacious rental flat in the center of Munich. This apartment has been our home for the 42 years since then. The quiet Kaulbach Street is within walking distance of the chemistry building, and the children's trip to school was even shorter. Moreover, art museums and theaters are close, and in the summers Trudl and I still ride our bikes regularly in the nearby English Garden, a city park larger than Central Park in New York.

In the mid-1950s many of the Kaulbach Street buildings still lay in ruins, some overgrown by greenery. Naturally, they were our daughters' favorite playgrounds. Our blood froze at the sight of Helga balancing high up on free-standing walls. This experience familiarized us with the clefts between theory and practice. As modern parents we knew that children have a right to their own negative experiences. Still, it was not easy to refrain from incessantly screaming "For heaven's sake, don't!"

"Daddy, are we so terribly poor? My classmates all have television at home; why don't we?" asked Helga at age 10. Listening to music while writing a manuscript or dining is regarded as a barbarian act by purists of the art; I plead guilty on that score. A TV program interferes with other activities and easily gains the upper hand. My countrymen spend a daily average of 3 hours in front of the TV screen. Admittedly, there are exciting programs on occasion; in 1968 we visited neighbors to watch the first man set foot on the moon. Our daughters today are as little infected by the TV bacillus as their parents.

Birge and Helga went through high school effortlessly. When I was visiting professor for 3 months at the University of Wisconsin in Madison (1959) and at Cornell University in Ithaca, New York (1962) it was no problem to take the whole family along. Initially Helga, then age 9, did not know a word of English. When we left Madison our

Captain's Dinner on the Ryndam; *the first transatlantic passage of the whole Huisgen family in 1959. From the left: Birge, 12; Trudl; Helga, 9; and me. On the invitation of W. S. Johnson, I served as Carl Schurz Professor at the University of Wisconsin. Also on board the* Ryndam *was a red Mercedes sports car for Bill Johnson.*

friends noted that Helga was the only one in the family speaking without an accent. That was not quite correct; Helga's intonation still bears the mark of the Midwest, although somewhat toned down at this point. Two years later Helga spent an additional year with Paul and Lou Bartlett in Massachusetts; their daughter Sarah was Helga's age. It was not an easy year for her, because she was unprepared for the clash of social customs. Helga's Munich high school did not give her credit for that year, because she had fallen behind in Latin, mathematics, and science. Helga's final verdict: American public high school is fun; German high school is hard work.

After completing the *Abitur*, Birge unerringly found her way to mathematics. She entered an academic career, specialized in algebra, and soon received professorships at Iowa City and Passau (Germany). In 1986 she joined the University of California at Santa Barbara. Birge followed her parents' example and chose her partner in her own field of specialty. Helga went into art history, then shifted to the study of Spanish and English. Her translations of fiction received wide acclaim. Helga, who lives in Munich, frequently inspects my English manuscripts, polishing style and grammar. Her standard reproach: "Dad, when will you accept it: no adverb between verb and object!" Helga presented us with a sweet granddaughter, Hanna, in 1990.

Munich, Our Love

Although we grew up in other parts of Germany, Trudl and I regard Bavaria, and Munich in particular, as our true home. Munich, Germany's "City of Art", has a flair of its own. Part of its personal imprint goes back to King Ludwig I, who reigned from 1825 to 1848, and his son Maximilian II (1848–1864) of the House of Wittelsbach, who were devoted to the arts and sciences. Ludwig I was responsible for the architectural enrichment of Bavaria's capital. Ludwigstrasse, Odeonsplatz, and parts of the Royal Residence breathe the spirit of Florentine Renaissance, whereas Königsplatz is a homage to classical Greece. Even though these styles are imitations, they are highly successful. Embedded in the cityscape, moreover, are pearls of Gothic, Renaissance, Baroque, and Rococo styles, the latter of course authentic. In the *Old* and *New Pinakothek* (art galleries) and the *Glyptothek* (sculpture museum), the art collections of the House of Wittelsbach are on display. On top of all this Munich displays a vibrant vitality, rather than resembling a museum recalling the past.

The university, founded in Ingolstadt in 1472, was subsequently moved to Landshut and finally transferred to Munich by Ludwig I in

1826. What was the reason for this odyssey? In the 15th and 16th centuries the students enjoyed so dubious a reputation that they were exiled to the boondocks so as not to endanger the virtue of the daughters of the nobility.

Over several postwar years an impressive hill grew in the north of Munich, made from the heaped-up debris and rubble of war destruction. The city elders of Munich restored and rebuilt its architectural splendor, a decision that I greatly value. No high-rise buildings are tolerated. The city's nickname, village with a million inhabitants, refers both to this architectural style and to the relaxed and informal atmosphere. Another saying, reflecting the Münchner's awareness of a privilege, holds a grain of truth: there are only two kinds of Germans, the ones who live in Munich and the ones who would like to live in Munich.

After the war Munich became Germany's cultural center, a role that may revert to Berlin in the future as a consequence of the reunification. Munich's theater scene, in particular, has been notable for decades. In the late 1940s Trudl and I traveled from Weilheim to Munich in overcrowded trains to see stage productions. We were spellbound by modern American plays that reflected much of our own life and emotions: Thornton Wilder's *By the Skin of Our Teeth* and *Our Town* or Arthur Miller's *Death of a Salesman*. Moreover, Bert Brecht's plays were frequently on the program, and so was classical Greek tragedy. The primitive stage settings of the time actually emphasized the actors' accomplishments and the intellectual substance of the scripts.

I have profited immensely from these assets of Munich, all the more because I regard theater, music, and art as a world complementary to that of science, with exposure to one acting as a stimulus for the other. Munich offers such a plethora of cultural events that careful selection is required, even for the cultural glutton. These days Trudl and I usually choose classical drama and music over the productions of the cultural avant-garde, probably because of our age.

My Brother Klaus and My Fascination with Art

My interest in art was originally kindled by my older brother, Klaus Peter, a highly gifted and creative painter and beginning poet. When Hitler raged against German expressionism, Klaus, then a high school student, never wavered in his admiration for the artists of *Der Blaue Reiter* (The Blue Rider) and *Die Brücke*, (The Bridge) to name just two major groups within the Expressionist movement of 1905–1925.

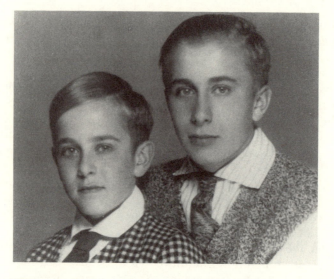

With my brother Klaus Huisgen (1917–1944) (right) about 1930.

Klaus enlisted at the age of 19. After completing his half-year of Reichsarbeitsdienst and the 2-year military service in 1939, he remained under arms. Hitler had just triggered World War II. Klaus served as a signal operator in the air force in Belgium, France, Russia, and Greece. On a furlough in 1941 he passed the entrance exam to the Art Academy in Munich, but was obliged to return to the front. Destiny denied him the chance of ever beginning a life in freedom and developing his talents. On returning from the Greek island of Rhodes, Klaus was killed in a plane crash in Styria (Austria) in the fall of 1944, a couple of months after the July attack on Hitler's life had failed.

Klaus was my closest friend, and letters had to fill the gap opened by our separation. For many years we exchanged our views by mail, and I still cherish the collection of letters I received from him. Moreover, drawings and watercolors reached me as his war diary. From his letters of 1943: "It is a frightful tragedy to see all that has been fashioned with the purest energy during many centuries, now destroyed with the means of our progressive technology." (March 18) "It is not timidity that makes one ardently long for an end to this harrowing war; it is the weight of a catastrophy which is drawing wider and wider circles." (April 17) "The gruesome development of this dreadful war makes one despair of humanity as the crown of creation." (July 16)

A corporal in a small military unit, Klaus had—astonishingly enough—friendly contact with the rural population of *the enemy* wher-

Klaus Huisgen (1917–1944): self-portrait as soldier, 1941; pen and black ink, ink wash. "How I crave the opportunity for peaceful and concentrated creative work." (from a letter of March 21, 1943).

ever he was stationed. After each move, the news that one of the German soldiers spoke Russian spread like wildfire through another Russian village. The sorrow of wartime during a Russian winter was depicted by Klaus as follows:

Müdgeweintes Land schläft ein
im weichen Flockenfall,
verloren treibt der Lieder Schall
aus mildem Hüttenschein.

Laternen schwanken hier und da,
von Schneekristallen hell umsprüht;
in jeder eine Hoffnung blüht.
Bald glühn sie fern, bald glühn sie nah.

Spuren wandern kreuz und quer,
durchirren müd die weisse Nacht,
der Winde Atem löscht sie sacht
und fegt die stummen Wege leer.

Klaus Huisgen, Russia 1943

A land tired out from weeping finds sleep,
On a cushion of snow it meets rest,
While songs forlorn from some human chest
Like mellow lights from the cottages seep.

Lanters swing in the wind that arose,
Enveloped by the crystal sprays,
Blooming each with hope as it sways,
Now glowing distant, now glowing close.

Tracks draw a pattern with crossings rife,
Erring, weary, through the white night's glare
Then are gently erased by the breath of the air
Which sweeps the soundless roads clear of life.

Translated by Birge Zimmermann-Huisgen (1993)

From a life that was not at all of his own choosing, Klaus escaped into the world of imagination. Writing poetry was his consolation. His death strengthened my resolve to concentrate on what I considered the essentials.

My interest in and occupation with art are passive; I am an avid admirer and reader, and a bit of a collector. I passed on the "virus" to my wife, Trudl, and later to our daughters during family visits to museums. There is an invisible barrier between admiring and first acquiring an original work of art. I was urged on and counseled by Wilhelm Foerst (1899–1986) who, as editor of *Angewandte Chemie,* promoted the journal to the status that it continues to hold. Foerst was a fervent collector of modern lithographs, etchings, and woodcuts; his judgment was based on an education of enviable range.

In 1957 I acquired the first woodcuts by Erich Heckel, followed by graphic art of Ernst Ludwig Kirchner, Emil Nolde, Karl Schmidt-Rottluff, Otto Mueller, Christian Rohlfs, and Max Beckmann. The subsequent continual contact with these pieces brought me closer to the evocative talent of these masters than I could have come through a glimpse in an exhibition. In my personal view, Expressionism is the most noteworthy period of German art since the late Renaissance, the time of Dürer, Grünewald, Cranach, Holbein, and Altdorfer.

Hitler denounced Expressionist art as un-German and ousted it from museums countrywide; the decried artists were forbidden to paint.

In my Munich flat in 1986, surrounded by 20th century art; my living habitat.

In 1937 the infamous exhibition *Entartete Kunst* ("Degenerate Art") was opened in Munich; at the same time, the pompous *Haus der Deutschen Kunst* ("House of German Art") opened its gates with a show of a strain of art (ironically labeled "blood and soil") that was touted as being close to the nation's heart. These simultaneous events drastically documented the catastrophe of having art defined by the government. When the full documentation of *Entartete Kunst* was published 50 years later, I was delighted to find myself thoroughly degenerate; I spotted a watercolor by Schmidt-Rottluff that is on display in the hall of our apartment.

Prices for Expressionist prints skyrocketed in the late 1960s, so I included other periods in my collection. George Braque and Pablo Picasso fascinated me. The latter, an incredibly productive and diversified genius, produced an abundance of graphic art. I am likewise fond of the abstract art of Ernst Wilhelm Nay (1902–1968), who is regarded as the greatest colorist in German art since Nolde. I feel that 20th century German art is still insufficiently known in the international forum.

African art caught my eye as well, because it had been a crucial stimulus for Expressionism, Fauvism, and Cubism. Only Trudl's protests limited the influx of sculptures of the Yoruba, Baule, and Lobi before our apartment became uninhabitable.

Marginalia on Art and Science

The worlds of art and science, both more or less elitist, share many points of contact. Perhaps my comments on connecting and differentiating features are neither profound nor novel, but they serve to outline my personal outlook on these two spheres. Is it wishful thinking to emphasize the *common qualities* of the two cultures in spite of the rifts and divergences? Notions like beauty, intuition, creativity, and abstraction are used in both art and science, though not always with the same connotation.

Most forms of art strive to produce images of the world, as perceived by our senses or imagination, by highlighting certain aspects and thus sublimating the phenomena to their essence. The complexity of the background is reduced and accidental elements are removed. The Japanese landscape garden, the ancestor figures of the Dogon tribe, the counterworld in Paul Klee's watercolors, or the verses by Reiner Maria Rilke epitomize this process of isolation and abstraction of an idea or sensual perception. According to Paul Klee, the artist thus makes the invisible visible.

Chemistry has to cope with a dizzying abundance of static and dynamic phenomena in the world of molecules. Chemical research aims at unveiling and ordering the factors in this intricate network, fitting them into a logical system. Arts and sciences both pursue the essential and the simple.

Both art and science moreover are founded on *creativity* and the power of *imagination*. In the popular view, the scientist watches nature and merely records observations. This is no longer true even for the descriptive sciences, let alone the experimental ones. Nobody can experiment with the universe, yet cosmology is making a fabulous dash forward in our time. Conjectures about the first seconds after the big bang reflect recent insights into the physics of elementary particles. In modern science theories and concepts frequently precede scrutinous observation and are subsequently substantiated or discarded.

How does science establish its trustworthiness? Although on several levels the cognitive process is not as straightforward as suggested, the role of imagination and intuition is uncontested.

When intuition joins exact research, the progress
of understanding will be accelerated astoundingly.
Paul Klee (1879–1940)

The successful development of chemistry in the 19th century relied to a certain extent on trial-and-error experimentation. By contrast, the contemporary experiment is triggered by increasingly complex questions, and its planning often requires unorthodox (i.e., creative) thinking. The experimental answer may confirm or reject a theoretical concept without obliterating the creative element of the original idea. This steady pruning of wild growth by comparing it with reality (or with what we take for reality) gives the development of science a *clear direction.*

The evolution of art, on the other hand, lacks such directive forces. During the period when art served only religion, it followed instructions; these rules, as well as general trends that were far more stringent than todays artistic boundaries, brought about some homogeneity. As a consequence, an art historian can pigeonhole an object of classic art according to its time and region of origin. Nevertheless the sequence of stylistic periods appears neither logical nor inevitable, whereas the traits of logic and inevitability are prominent in the history of science. The paintings in the cave of Lascaux (~20,000 years ago) conjuring up hunter's luck seem more modern to us than Claude Lorrain's (~1650) and Constable's landscapes (~1820). In the Egyptian sculpture of the Middle Kingdom, the squatting figure was boldly reduced to a stone cube from which the head protruded. In my opinion this block statue (~2000 B. C.) complies more with present art concepts than the doryphoros (spear bearer) of Polyclet (Classic Greece, ~450 B.C.) or the Venus de Milo (~100 A. D.).

Can we recognize any progress, or at least a unified direction, in our avant-garde art? I personally find it hard. Originality has become an obsession; modern artists are eager to undermine any definition of art. Is "anything goes" chaos or liberation?

Beauty is a fuzzy term in both of these two cultures. In art it originally referred to harmony of colors and shapes. To be perceived as beautiful, the representation of the human figure, for instance, had to obey a certain canon. In many cultures, beauty was connected with *symmetry*, an archetype of human thinking. It is richly documented in the art of Old Egypt, embracing architecture, sculpture, and painting. In the Tombs of the Nobles at Thebes, I saw frescoed double scenes showing the deceased hunting (~1400 B. C.). The artist slightly disturbed the mirror image, in obvious awareness of the fact that rigid symmetry tends to diminish liveliness and aesthetic appeal.

> *If one could explain a picture, it would no longer be a*
> *work of art. Shall I tell you which qualities I think are*
> *important for real art? It has to be indescribable and*
> *inimitable.*
>
> Pierre-Auguste Renoir (1841–1919)

In contemporary art history, beauty is no longer regarded as a useful term. What does beauty stand for in science then? I suppose when Einstein or Heisenberg mused about the beauty of physical theories, they were referring to unity, consonance, scope, and simplicity.

The *beauty of molecules* in the chemist's language most often refers to symmetry. Personally, I confess to a certain aesthetic drive behind my choice of pentazole or cyclooctatetraene as objects of study. I felt fascinated by the dynamics of electrocyclic reactions (page 125) in which a plane or twofold axis of symmetry is retained. In a splendid essay Roald Hoffmann[607] reflected on *Molecular Beauty*, considering phenomena as crystal structure, novelty, complexity, and the rainbow of biological activities. *Elegance* is another aesthetic notion. Although it is not important as an epithet in art, the chemist speaks of an elegant synthesis or of elegance in establishing a structure or a mechanism. Here elegance usually stands for power of persuasion and unexpected simplicity.

Beauty and elegance likewise reign in mathematics, the purest of humanities in that all its objects, methods, and conclusions are offspring of the human mind. Pythagoras and Thales derived their geometrical principles from a form of intuitive logic, undoubtedly motivated by the subject's aesthetic spell.

The "Art of the Fugue", the last and quintessential opus of Johann Sebastian Bach, suggests a mathematical scheme in its rhythmic and melodic variations. The worlds of music and mathematics appear to merge in this masterpiece.

In contrast to science, the artistic intuition giving rise to an opus does not require subsequent validation. However, it is nonetheless worth discussing whether an idea is adequately expressed in a given artistic work. The concept of *quality* is omnipresent in assessing the work of artists as well as of scientists. Except for the achievements of a few pioneers wildly ahead of their time, scientific merits are usually recognized within their communities. Although quality does not have a rigorous definition, there is a high degree of consensus among scientists.

Some creative artistic giants like Michelangelo, Rembrandt, and Beethoven stood out more or less undisputed from the start. Nevertheless, there are no unambiguous criteria of artistic quality. Particularly in modern art, two people—artists included—will rarely agree in the assessment of quality. Where I shrug my shoulders, someone else will speak of a masterpiece.

Superimposed on individual fluctuations of taste are worldwide trends, which today are further promoted by media and exhibitions. Such changes in the evaluation of art are by no means novel. During the long construction time of European cathedrals, stylistic preferences often changed, for example, from the austerity of the Gothic period to the elegance of the Renaissance, continuing from there to the opulence of the Baroque period.

Not even the personal evaluation of art is immutable. Open interaction with a variable ambience modifies our own standards. Thirty years ago I ridiculed the exuberance of *Jugendstil* (art nouveau). Later my eyes were opened by exhibitions, essays, and architectural masterpieces. By the time Richard Riemerschmid had completed the Kammerspiele, one of Munich's theaters, in 1912, the public taste had turned away from *Jugenstil*, a tragic development for the great architect. His work was essentially covered up with plaster. In 1970, however, the art nouveau interior of the Kammerspiele was restored to its original splendor.

Market forces rule in both art and science. Unless financially independent, the artist must somehow adapt to the predilections of paying clientele. When research money dries up, many scientists work on programs favored by government agencies. They learn to play the game of the correct keywords; in both organic and inorganic chemistry, the prefix "bio" and allusions to "new materials" have proved very useful.

Why do art and science give us pleasure? The message of the artist plays on the whole range of human emotions, from joy to grief and from elevation to compassion. Science, on the other hand, appeals to a narrower section of the human mind, that governed by rationality. But this rational portion includes curiosity, one of the strongest human drives. Satisfied curiosity is one of my foremost gratifications—at least until the next problem comes up.

On Women in Chemistry

It was not until many years after the fact that I began to fully respect and value the sacrifice that my wife, Trudl, had made in giving up her career as a chemist. In the 1930s only a small number of chemistry students in Germany were female, and their professional chances were scant. Working at the laboratory bench for a lifetime was considered too strenuous for women. Industry hired women chemists mainly for service in libraries and patent offices. The research director of a big company once—probably after a drink—gave me his blunt opinion about the role of women in the chemical industry: "If she is pretty, one of our male chemists will fall in love with her, and both will be distracted from work. If she is plain, she will turn sour in her forties, and that is even worse."

One of the few women with successful academic careers in German chemistry was Elisabeth Dane (1903–1984). She assisted Heinrich Wieland in his research on bile acids, for which he was honored with the Nobel Prize in 1928. Dane became an associate professor at the Munich Institute; for several decades she trained medical students in chemistry and did notable independent research work. Moreover, during 1945–1952 she was Wieland's aide and consultant in the running of the institute. During my first years as a professor in Munich, I often benefited from Elisabeth Dane's advice.

Of course, the conditions facing women in chemistry have changed fundamentally for the postwar baby-boom generation. The percentage of women among beginning chemistry students in Germany rose from 12% in 1960 to 33% in 1990. Interestingly, the fraction is lower in the group of Ph.D. students, 4% in 1960 and 20% in 1990.[608] As far as I can see, female chemists are no longer disadvantaged professionally.

Was I biased against female graduate students? Bad luck on my part; I just did not attract more than two.

On Academic Customs

Institutions of higher education are bound to tradition and proud of their age. Like most young professors, I started out by ridiculing such academic customs as the wearing of gowns at official celebrations. I learned to appreciate the worth and weight of tradition only when it was jeopardized by radical students in 1969–1973 (page 13). Student functionaries stirred the masses to demonstrations against capitalism in general and an obsolete university system in particular. Their fantasy university was controlled by the students, and exams were disdained as below human dignity.

I joined the *Bund Freiheit der Wissenschaft* (Association for the Freedom of Science) as soon as it was founded in 1970 to ward off any dangerous landslide in our education system. By manipulated elections the "Red Cells" had achieved control of the official student representation at the major universities, including Munich. One morning in 1971 the cafeteria of our institute was decorated with banners expressing sympathy with the Bader–Meinhof gang, a terrorist organization. The wavering majority of our students certainly did not approve. Afraid of sabotage, we equipped the spectrometer labs with steel doors.

My attempts to save our chemistry courses and to keep the discussion rational sapped both my time and energy. Perhaps many of my colleagues who avoided confrontation were wise in doing so. Finally the student revolution, with strikes and disrupted lectures, petered out on its own. Did it really? The professors have not resumed marching in gowns, although I would now welcome the renewal of this tradition.

Opposition of young people against the establishment is part of the moving force of progress. The worldwide student protests in our time may have been an extreme reaction to unsettling new ideas of personal freedom and to the overwhelming complexity of modern civilization. Rethinking of old customs may have been necessary but, as in every mass movement, the clamorous elements were not the sensible ones.

My relation to tradition is ambiguous, insofar as this term may disguise inertia and rigidity. Curricula should be flexible enough to adapt to the progress of science and changing priorities. In order not to be overwhelmed by whims and short-lived fluctuations, universities often must steer a wise middle course. Popular or not, striving for a high level of achievement should remain an academic tradition.

On My Young Associates

Many hands must work together to verify new concepts experimentally and to test their fruitfulness. Cooperating with generations of students has been an exhilarating experience for me. I was blessed with many an excellent student and postdoctoral fellow, and my gratitude is sincere.

Efficiency in research is a complex function of many variables; this rule applies to professors as well as students. Again and again, I have been amazed at the wide gamut of talents, temperaments, and skills among students.

When a new student wanted to join the group I usually provided a choice of topics, often written for each project. The writing forced me to thoroughly ponder the problem. The frequency of my visits at the bench and of discussions of the results varied, of course, with the efficiency of the co-worker and with my interest in the problem at the time. Two lab technicians would be assigned to the very best students, although the less brilliant ones might have been more in need of help.

These technicians were usually young women, and many of them married graduate students—more proof of the binding forces of chemistry. My feelings were ambiguous because in each case I had to find and train another technician.

I asked my young associates for bimonthly reports; it was often sheer drudgery to get them. Writing about results makes the student aware of experimental shortcomings and triggers an analysis of possible consequences. Whenever feasible the report formed the basis for a discussion, often of many hours.

My first postdoctoral fellows arrived in 1954, and most of them came from the United States. Around 1970 the stream fizzled; I learned that a fellowship within the United States is likely to be a bigger boost for careers. Postdoctoral fellows from Austria, England, France, Italy, and Spain, as well as from Australia, China, India, and Japan took their place. The Alexander von Humboldt Foundation usually provided support and even penetrated the Iron Curtain. Thus, I cooperated with

My birthday party at the Institute in Munich in 1963, one of the social highlights of the year. From left: Trudl, ?, me, Klaus Hafner, Joseph F. Bunnett, and Christoph Rüchardt. It was the occasion for presenting the photo chronicle of the past year, spiced with impish comments and photo montages.

young chemists from Czechoslovakia, Hungary, Poland, and Russia. Politics interfered only once; the foundation had extended the stipend of a young chemist from Leningrad (now St. Petersburg again) in 1984, but the Russian Embassy sent him home.

Visiting professors and postdoctoral fellows from foreign countries contributed to the scientific atmosphere, the international flair, and the social life of the "lab family". Joint skiing excursions and parties for new arrivals and fresh doctorates promoted personal contacts.

A student joke at a birthday party in Munich in 1965 was an exhibit of stamps from the "Famous Germans" series. It roused roaring laughter from all, including the one portrayed.

What is the optimal size of academic research groups? I disliked the "officer system" and preferred directing and discussing without middlemen. In the early 1960s my group grew to 24 and I had to ask my associate Rudolf Grashey (page 93) to help by supervising the 1,3-dipolar cycloaddition team. I often succeeded in persuading students to stay beyond their doctorate for a year or two; they took care of the beginning graduate students when I was abroad. Apart from the two technicians and a secretary, no permanent staff member worked for me.

I felt relieved when the group size fell below 20 again; in my experience the total output of research results levels off at 15 associates. I marvel at colleagues who can handle 40 and more co-workers. Over the past 10 years the number of my graduate students has decreased, and I have worked mainly with qualified postdoctoral fellows. I no longer feel adapted to the changing social climate, in which the scientific engagement of most students is diminishing. Possibly, my demands and expectations have increased.

So far I have directed the diploma and Ph.D. theses of about 150 students; 75 postdoctoral fellows joined the effort. From each of the two camps, graduate students and postdoctoral fellows, more than 20 are now holding academic positions themselves. Thus, I have had the pleasure and privilege of founding a school.

When I became professor emeritus in October 1988, I reveled in pleasant reminiscences generated by a book, *Rolf Huisgen und sein Arbeitskreis* (Rolf Huisgen and His Group), which was presented to me at a get-together of my former and present associates. Gernot Boche of Marburg had asked former graduate students and postdoctoral fellows for their memories of the Munich Institute, and many of them responded. Boche combined the reports into a book illustrated with photos and supplemented by lively and witty articles by friends and colleagues, a true token of empathy.

The editor of this series asked me to comment on the methods of conducting "efficient research". I wish there were a general recipe for reaching this goal. I believe that very bright students need only experimental guidance, and the less gifted ones will not make a big impact on science in any case. The best method (tongue-in-cheek) is to ascertain in advance that a student is capable, before deciding on acceptance for graduate work. I spoke with students, checked their exam reports, and studied their records in the lab courses. As a result, I have been lucky very often. Before inviting postdoctoral associates to join my group, I tried to get the relevant information, including knowledge about the candidates' skills. I asked the doctoral program directors not to send me "wooden nickels."

Should graduate students be kept on a short or long tether? It depends on the student. Young graduate students usually need extensive advice on how to plan and carry out an experiment so as to make it release a maximum of information. Students should be taught to grow "yield-oriented" and to check each operation by weight. I tried to convince my co-workers that a good melting point is a better test of purity than an X-ray analysis and that a stack of spectra is no substitute for correct elemental analyses. I stressed the importance of quantitative NMR analysis with weighed standard. (A list of standards is kept on the bulletin board.)

The effective techniques for motivating young associates are quite limited, I believe. The professor's own level of enthusiasm is, of course, essential. In the end, however, most of the motivation and enthusiasm must come from the student.

Striving for excellent grades—these being prerequisites for a successful career—is a legitimate component of this drive. Of course, external circumstances may either foster or squelch such determination. Our high standard of living in the Western world offers an overdose of distractions, against which only dedicated students are immune. In this respect the postwar generation of German scientists actually had an advantage. The challenge of rebuilding Germany from the ashes, during the 1950s and 1960s, provided a massive boost to the motivation of young scientists of that era.

Notes on Publishing, Lecturing, and Consulting

Several of my co-workers' Ph.D. theses of the late 1950s and the 1960s are still awaiting publication; some results were merely mentioned in reviews. Typically, when my graduate students gave me drafts of their theses, I invested a great deal of time in counseling them and revising their expositions. When my research areas shifted, thesis drafts of latecomers would still be carefully revised, but the composition of the final paper would be delayed. Unfortunately, publishing did not keep pace with production.

Neither precipitate publishing nor decades of procrastination serve the scientific community, but sometimes it is not easy to find the golden balance of speed and thoroughness. I waited until experimental evidence was conclusive, accepting the risk of being scooped by others. Although my bibliography exceeds 500 items, there is quite a backlog of papers yet to be written. This burden bothers me because I feel a strong obligation toward my co-workers to publish the results of their endeavors. My perfectionism, apparently incurable, aggravates the situation, although I am quite familiar with the *principle of diminishing returns* on revising papers. Why not try doing a little less than my best? *Good intentions are futile efforts of overriding the laws of nature.* (Oscar Wilde)

My striving for a personal style in writing and speaking may appear old-fashioned today; it is a time-consuming hobby as well. Linguistic proficiency and elegance of style are no longer valued as highly as they were when I received my education. In the 1960s one of our state ministers of education coined the phrase, "High German is the dialect of the ruling class." A linguistic decline, especially among students, is widespread throughout the scientific community.

To me language is more than a service vehicle for transporting scientific ideas; it is an aesthetically motivated extension of the fun connected with scientific research. I therefore regret not having learned

English in school. I studied the language independently as an adult and never attained a versatility comparable to that which I have in my mother tongue.

There are as many ways of writing a paper as there are individuals. Having the first draft typed, corrected, etc., in several rounds is a frequently recommended procedure that has been made efficient by the computer. However, I cannot give up my old-fashioned method of toiling with pencil and eraser, sometimes with scissors and glue. I justify this messy procedure with the fact that it requires only one round of typing unless excess length needs to be reduced.

Originally I was reluctant to publish preliminary communications, but I soon realized that they are more widely read than full reports. I now find short papers and reviews extremely helpful in coping with the flood of information. Although I have written all short communications in English since the late 1960s, I still compose full papers in German. However, I am painfully aware that the use of my native language is becoming increasingly problematic because of the predominance of English in modern science.

The rabbitlike proliferation of highly specialized journals during the past three decades has become an impossible strain on the budget of scientific libraries. In 1973 I participated in a public appeal entitled "Too many chemistry journals", which was initiated by Roald Hoffmann, but the proliferation was not halted. Most of the new periodicals are superfluous in my eyes and will not be as widely read as the traditional ones. Important new developments might have a slimmer chance of dissemination when hidden in a specialized journal.

On the other hand, I do not object in principle to a diversity of journals. To a certain degree, the varying quality of periodicals prioritizes routine reading because when pressed for time, we can concentrate on the more important publications. In this respect even the journals at the low end of the scale serve a function. What bothers me is that many of the new specialized journals claim to being of the highest quality right from the outset, before natural selection has sorted things out.

I consider lectures to be superior to secondary literature as a tool for spreading information because the personality of the lecturer is more directly involved. Most good scientists are also good lecturers; I know of only a few exceptions. This statement, of course, depends on my definition of good in this context. Unlike the politician, the scientist is not required to study rhetoric, but being familiar with a few didactic principles does not hurt. A well-structured discourse, delivered with verve and eloquence, helps in stimulating the audience; a dash of entertainment may be a strategic aid in keeping it awake.

Listening to a lecture on unfinished work often has the charm of personally attending the emergence of a new idea. As for myself, I have always preferred to report on completed, even polished, chapters of research. I like to include mechanistic criteria to answer anticipated objections.

Formal versus informal lectures: the former are more in accordance with European tradition. In the early 1950s I was surprised to see some American lecturers perched on the speaker's desk chatting amiably. I have no objection if the chat is well-founded and well-presented. On the other hand, I consider poorly prepared lectures a discourtesy, showing disdain for the audience. But then, after all these words, the essence of a good talk is simple: excellent chemistry. Not even inspired volubility can make up for a lack of substance.

Warm and friendly relations between university and chemical industry are a German tradition. The *Fonds der Chemischen Industrie* (Fund of the Chemical Industry; page 5) maintains its beneficial activity, even during the 1993 period of economic recession. Furthermore, the big chemical companies have a lecture program in which university professors report on the progress of their basic research. On the other

With Horst Pommer (1919–1987) (left) before my lecture at BASF AG in Ludwigshafen in 1977. In the background are several of my former students. As director of research at the BASF AG, Pommer splendidly adapted the Wittig olefination to the industrial syntheses of vitamin A and β-carotene. (Photograph courtesy of BASF AG.)

hand, leading industrial chemists are welcomed as guest speakers at university seminars. Many chemical industry research directors teach as *Honorarprofessoren* (honorary professors) on a more or less regular basis at a university of their choice; students are delighted with this first-hand information.

Consulting for the chemical industry plays less of a role for the German professor than for his colleagues in the United States and some other countries. I myself had friendly relations with the pharmaceutical research department of the Hoechst AG since my beginnings as *Privat-dozent* (lecturer) in the laboratory at Weilheim (page 5). Gustav Ehrhart (the discoverer of methadone) was succeeded by Heinrich Ruschig and Günther Seidl as laboratory directors; I admired their capacity for mastering chemistry, biology, and medicine, three fields that have grossly increased in complexity. Günther Seidl, my former associate in the chemistry of medium-sized rings (page 43), asked me to consult on a regular basis for the pharmaceutical synthesis; the Hoechst AG has the largest pharmaceutical production in Europe. I learned much in these discussions (1972–1989), and my eyes were opened to the vexing problem of biological activity.

On Scientific Contacts and Travels

Common research interests may generate durable bonding forces and spawn friendships. Although we are simply born into a nation, with no input from our side, our fields of scholarship reflect personal choices. It is therefore hardly surprising that, both in my own work and in my view of science at large, I have been strongly influenced by ties among scientists that span borders and continents.

Traveling is one of the great bonuses in the active scientist's life, even though it may lead to "travelitis", an infectious disease. Whether I am infected, I cannot quite decide. At any rate, I was exposed to the bacillus and enjoyed this exposure; I still do. Aside from the scientific contacts involved, I love learning about foreign countries, enjoying scenic beauty, and visiting places of archeological interest.

"Travelitis" on the part of American scientists was a benefit for me in Munich. They stoked my enthusiasm for reaction mechanisms. As a consequence, contacts with American chemists were of prime importance for my work. The Rockefeller Foundation kindly offered me a travel fellowship in the early 1950s. When the construction of the new chemistry building in Munich was nearly completed in 1955, I submitted my travel plan involving 26 laboratories in the United States. Gerard R. Pomerat, one of the directors of the foundation, insisted in his letter of approval that I include weekends in national parks on the way to and from the West. On this trip my wife and I fell in love with the desert scenery of the American West, which has no equal in Europe. We were to return many times.

Visiting professorships offer a multitude of blessings: exposure to a new scientific atmosphere, escape from routine chores, and, as a consequence, more leisure for thinking and writing. In 1959 I spent 3 months at the University of Wisconsin, where William S. Johnson, Harlan Goering, and Gene van Tamelen were my main discussion partners. We stayed in a "real American house", a major thrill for our children. Sam and Helen McElvain had invited us to use their home and car while they were abroad. I guess in some ways these were the "good old times" for the United States, too; our neighbors in Madison were amused about our locking the front door at night—an unheard-of degree of paranoia. As a result of my visit, in 1956 the faculty at the University of Wisconsin, Madison, offered me a full professorship there. This opportunity presented me with a tough dilemma—my cooperation with excellent students was my main reason for remaining in Munich.

As a Baker Lecturer at Cornell University (1962) I enjoyed contacts with Jerry Meinwald, Charles Wilcox, and M. J. Goldstein. Several times I had luncheon with Peter Debye, who, at the age of 78, dealt with electrostatic phenomena in flowing liquids. As an emeritus (and Nobel Laureate of 1936), he had *to rent* the lab space. Harsh customs, indeed, were prevailing at the time.

During my stay at Cornell I wrote review articles on 1,3-dipolar cycloadditions[159,300] and recognized the 1,3-dipoles of Chart V as hetero derivatives of the allyl anion. Thinking in terms of MO theory was not as common for an organic chemist then as it is today. However, the simple idea stood up to scrutiny when I discussed the experimental evidence[300] with my colleagues. As an extra bonus, it also clarified the relation to the Diels—Alder reaction.

The Pacific Coast Lectureship, coordinated by Syntex Research, was a splendid opportunity for me to visit the major chemistry departments between Los Angeles and Vancouver (British Columbia, Canada). In 1968 Trudl and I followed the spring bloom from south to north. The scientific program was coupled with scenic adventures such as Anza Borrego Park in California and Crater Lake in Oregon. Rainbow Bridge in Utah and the Grand Canyon with its Kaibab Trail to the Colorado River were the grand finale.

On the Berkeley Campus, in 1968, we watched students preparing banners with strike slogans as a matter of routine. The rebels' attentions were, however, diverted by a fellow student's skill in body-painting a bikini beauty, yet another fad of the time. They only reluctantly joined the daily protest march in the bright sunshine.

The summer climate in Florida is merciless, but fall is lovely to

With Trudl at Easter 1968 in Anza Borrego Park, California. The excursion was an adventure, with rattlesnakes, ocotillo trees, and an earthquake at 4 a.m. (Photograph by J. D. Roberts.)

make up for it. Paul Tarrant was my host at the University of Florida in Gainesville in the fall quarter of 1968. Moreover, I shared interests in diazo and carbene chemistry with William M. Jones. On the side, Trudl and I learned what a hurricane is all about. "Good thing we brought our rubber boots", we said as the water rose to our knees. (This is only slightly exaggerated.)

Three times during the 1970s I gave advanced courses at the University of California at Davis. My scientific discussions with Ray Keefer, Al Bottini, George Zweifel, and others were delightful. Short visits to the other branches of the University of California proved highly gratifying as well. The campus of UC–Davis is located in an arboretum, and spring comes early. To us, the almond blossoms during the second half of February were like a dream. Davis is known as the town with a record number of cyclists. We did not verify this claim, but quickly found that getting on bikes ourselves was the only mode of survival. We still correspond with our friends in Davis.

On shorter trips I served as Welsh Foundation Lecturer (1960), Max Tishler Lecturer at Harvard (1963), Morris Kharasch Lecturer at the University of Chicago (1974), and other similar assignments. Beyond pleasant scientific contacts, we found that Jack and Helen Halpern in Chicago were art buffs quite akin to us. Trudl accompanied me only on longer visits; she did not much care for the restlessness of packing and unpacking on trips with many stops.

At the Gordon Conference on Heterocyclic Compounds in New Hampshire in 1967. Who can resist the temptation when Ted Taylor (left) leads an afternoon hike? My sport shirt was a California acquisition; I usually went on hikes in an outfit designed for Main Street.

I have always envied people who never hurry and who have full command of their time. Time pressure has been my permanent companion over the years. Consequently, I used to attend congresses and symposia only when I was invited as a lecturer, gathering as much information as possible on such occasions. In particular, I liked the Gordon Conference in New England and the more luxurious versions, the Bürgenstock Conferences on Stereochemistry in Switzerland. On the

At the Bürgenstock Stereochemistry Conference in 1982. With Jack E. Baldwin (right) anticipating a lecture.

25th Anniversary of the Bürgenstock Conference in 1989. On the left is André Dreiding, founder of the Conferences. I am contributing to the discussion. (Photograph courtesy of K. Zimmerman.)

other hand, the "symposium of two" (the one-on-one discussion) remained more fruitful for me.

Not all of our travel routes took us west. Eiji Ochiai, Tokyo, broke the ground in the chemistry of aromatic N-oxides.[609] In 1962 he suggested that Ken-ichi Takeda, director of research of the pharmaceutical company Shionogi, invite me as speaker for the inauguration of their new research laboratories. Discussions with Shionogi chemists, visits to several universities, and the highlights of the classical temple and garden architecture of Kyoto and Nara made up a compact 2-week program.

Ochiai had spent a year in Frankfurt as a graduate student and published in German. He was a wonderful host and interlocutor, and the reverence of his former students—many of them in academic positions—for their old master did not surprise me.[610]

Arrival at the new research laboratories of Shionogi and Company, Ltd., in Osaka, Japan, in 1962. From left: Ken-ichi Takeda (1907–1991), Eiji Ochiai (1898–1974), K. Nakagawa, and me.

When Ochiai heard that his German visitor was interested in Ukiyo-e, the colored woodcut marking a great period of Japanese art (1650–1850), he telephoned the director of the graphic collection in the National Museum. "Yes, the foreign visitor may see the early prints of Hokusai's *Thirty-six Views of Mount Fuji* tomorrow if it does not rain."

Luckily, it did not, and the valuable pieces were taken out of their air-conditioned safe.

The cultural clash, in particular the struggle between the highly developed Japanese aesthetics and the more pragmatic Western way of life, became an intergenerational problem to some extent. My hosts spoke of a peaceful coexistence of the two styles of living; one devouring the other appeared like a more fitting description to me. When I revisited Japan in 1981, Westernization had made headway. The concert world is dominated by European composers, and the beautiful airs of koto and shamisen more or less rank as folk art.

With Teruaki Mukaiyama (left), University of Tokyo, on a visit to Munich in 1979. This scientist is one of the great masters of organic synthesis who developed new methods, such as redox condensation.

On my second visit to Japan (1981), this time with Trudl, I was a fellow of the Japan Society for the Promotion of Science, and Teruaki Mukaiyama was my host. The hospitality was again overwhelming. My former Japanese postdoctoral fellows—most of them now professors—gave us a cordial welcome, too.

In its struggle over the Equal Rights Amendment the United States may have been ahead of the European countries, but in 1981 Japan was even further behind than Europe. Trudl was invited to come along for dinner after my lectures, but often found herself the only woman at the table. One evening at Sendai, Trudl was on the verge of

Akihiro Ohta (left) showing the beauties of Kosokuji Garden in Kamakura to Trudl and me (1981). A former student of Ochiai, Ohta dealt with cycloadditions of diazoalkanes as a postdoctoral fellow in Munich (1965–1966). A professor at the Tokyo College of Pharmacy since 1971, Ohta studies topics such as pyrazine chemistry and Pd-catalyzed substitutions.

bowing out; she changed her mind when the wife of a colleague arrived to pick us up. However, the lady only dropped us off at a restaurant, and the situation of the previous evenings was repeated.

Japanese science rose to high standards unusually fast. The diligence, creativity, and efficiency of Japanese scientists indeed call for respect. In chemical research, associate and assistant professors usually cooperate with full professors. I regret this custom; the series of names on Japanese publications often gives no clue as to who led the way. In my opinion, the field as a whole profits when young scientists are given an early chance to start independent research.

An exchange of Chinese and German scientists was arranged by the *Academia Sinica* (Chinese Academy of Sciences) and the *Max-Planck-Gesellschaft* (Max Planck Society). My two Chinese postdoctoral fellows were competent and efficient although, because of the dreadful Cultural Revolution, they had next to no research publications at the age of 40. Formally, they were not even postdoctoral fellows because all titles had been abandoned in China. They have been reintroduced since then.

When I visited China in 1982, Wang Yu, the venerable director of the Shanghai Institute of Organic Chemistry (SIOC), was my host. Wang had received his Ph.D. with Heinrich Wieland in Munich in 1937. In 1965, the news of the total synthesis of crystalline pure insulin hit Western science like a bombshell. It was Wang's achievement. Later he directed the cooperation of several Chinese institutes in yet another synthesis of international rank. The transfer-RNA of yeast for alanine with all its 76 ribonucleotides and odd bases was the result of 13 years of back-breaking endeavor. Research accomplishments in countries without a tradition in science and not backed by an efficient chemical industry deserve special admiration. For the same reason, the research work of the late Costin Nenitzescu (1902–1970) in Bucharest, Rumania, impressed me strongly.

In 1982 the Chinese Chemical Society celebrated its 50th anniversary in Nanjing, and I presented the address of the *Gesellschaft Deutscher Chemiker* (German Chemical Society). Wang Yu gave the main lecture, dealing with the aforementioned synthesis of a *t*-RNA. He concluded the meeting with a poem praising the beauty of Nanjing, located between the Yangtse River and the Hsuan-Wu Lake with its five picturesque islands.

The Munich Faculty customarily renews the doctoral diploma of prominent scientists after 50 years. Wang, at the age of 77, received this honorary diploma in 1987 and lectured for the occasion on despeptidoribonucleases.

On revisiting China in 1993, I found that Westernization could not be overlooked; Beijing with its numerous skyscrapers appeared as a New Chicago. Wang Yu at the SIOC offered a hearty welcome; at age 83, he still spends several days per week in the laboratory. I enjoyed discussions with Jiang Xi-Kui, an excellent physical organic chemist, and his group. My English lecture was not translated sentence for sentence into Chinese as on my first visit. However, I was not certain whether the majority of students and young staff members fully grasped the spoken words; clear slides were helpful.

On my first China visit, I was invited to see the awesome clay army in the Tomb of Qin Shihuang (first emperor of Qin, 210 B.C.) as well as the Tang Tombs at Qianling, both places neat Xi'an. In 1993 a weekend excursion led to Hangzhou, the capital of the Southern Song

Arrival at the Shanghai Institute of Organic Chemistry, Chinese Academy of Sciences, in May 1993. From left: Lin Guo-Qiang (director), Wang Yu, me, and Jiang Xi-Kui. The outstanding work of Wang Yu turned this institute into a Mecca of organic chemistry in China.

Dynasty, located at Xihu (West Lake). The scenic and cultural attractions around Xihu currently draw masses of tourists, mainly Chinese ones; tourism is among the newly gained freedoms.

On the same 1993 trip I visited South Korea on invitation of Yung-Bog Chae, my former graduate student (Ph.D. 1965) and now President of the Korean Research Institute of Chemical Technology at Taejon. My first lecture at the Seogang University, Seoul, was scheduled for Saturday, 5 p.m.—no misprint; the institute was bristling with life on a Saturday afternoon. Soon I realized why the economy of this country is still rapidly growing.

South Korea was repeatedly ravaged by wars, the last time in 1950 by North Korea. Historical monuments, so important for the national identity, have been rebuilt; presently the architectural heritage of the old Silla Empire is under construction at Kyongju.

Although relations between Israel and Germany are naturally burdened by the past, scientists may help to mend the severed bonds. I was delighted when Gerhard Schmidt (1919–1971) of the Weizmann Institute invited me in 1971 to tour Israel's universities. Schmidt, an X-ray crystallographer, pioneered in solid-state photoreactions. On a weekend excursion, he showed Hans Schmid (1916–1976) of Zurich, another guest at the time, and me the ruins of Avdat, a Nabatean settlement lost in the Negev desert, and the Herodian fortress of Masada, site of a tragic and heroic chapter of Jewish history. We also saw an achievement of the modern Israel. Mountains of glaring crystals of KCl and a whole pallette of chemicals at the Dead Sea Works result from novel isolation and separation procedures. Gerhard Schmidt showed the mark of illness; some months later he died on a lecture trip to Basel.

In Haifa, Israel, in 1980. With Hemdah Ginsburg (1920–1990), David Ginsburg (1920–1988), and Trudl. David Ginsburg was one of the most influential scientists in the young state of Israel.

In the late 1970s my close contact with David Ginsburg of Haifa led to a friendship between the two families. In 1980 I spent 6 weeks at the Technion, Haifa, lecturing on cycloadditions. During our stay Trudl and I were invited to participate in the Passover celebration, an event I am particularly fond of recalling. The Ginsburgs, in turn, spent the summer semester of 1986 in Munich to share our chemical and cultural life here. David Ginsburg is best known for his total synthesis of morphine and the extended studies of propellanes. He passed away in 1988; I still feel the loss.[611]

Ginsburg was a great storyteller. His anecdotes of meetings with renowned chemists were captivating. Moreover, the born New Yorker spoke a highly cultivated English, which tended to whet his tongue even further. His "Plea for a Renaissance of a Humanistic Style in Scientific Papers" heading Vol. 1 of the *Nouveau Journal de Chimie* (1977) asks:

> *Must scientific progress bring in its tempestuous wake a meagre*
> *style characterized by scanty vocabulary, poverty of phrase, con-*
> *fined by most editors to routine strait-jacketed expression?*

In 1987 I accepted an invitation of the Israel Academy to serve as Einstein Visiting Professor at the Hebrew University of Jerusalem. The visit brought new contacts and new friends; shortly afterward Joseph Klein came to Munich for several months.

I will record only a few of my travels within Europe. In 1961 I was Centenary Lecturer of the Chemical Society (London) and often had the chance of meeting English colleagues. Several times I gave lectures and courses in Spain; in 1972 I received my first honorary degree from the *Universidad Complutense*, Madrid. Less frequent were trips to the Eastern countries when they were still under socialist rule: German Democratic Republic, Czechoslovakia, Hungary, Poland, and Russia. I was welcomed cordially, but signs of discouragement on the part of my colleagues were not rare. In Leningrad (1984) I met Irina K. Korobizina (1919–1990) again. On her visit to Munich in 1970 she had expressed a wish for a discussion with our "socialist" student rebels; afterward, visibly shaken, she noted that our students were confounding socialism with anarchy.

At a Leningrad hotel in 1984 I noticed with a somewhat queasy feeling that opera and concert tickets were much more readily available to West German visitors (who paid in Deutsche Marks) than to the citizens of socialist East Germany. On a visit to Hungary in 1978 my wife and I attended a performance of Verdi's opera *Nabucco* in Budapest; the hymn to freedom of the oppressed Hebrews aroused a standing ovation.

Poland, for centuries an underdog among powerful neighbors, held my sympathy. In 1981 I visited institutes of the Polish Academy of Sciences in Warszaw and Lodz; their research was very good, but the suppression of "Solidarnosc" formed a macabre background. Recent visits in 1987 and 1992 included the University of Lodz, as well as those of Wroclaw (Breslau) and Gliwice (Gleiwitz). The young democracy is still steering through turbulance, but the economic indicators are pointing upward.

When I revisited Leningrad in 1992, the beautiful city, Russia's former capital at the Gulf of Bothnia, had regained its old name of St. Petersburg. In the research discussions at the technological institute and the university, the dedication of the Russian colleagues to their work amid political turmoil and with the economy in shambles impressed me as much as their hospitality.

Another lasting friendship developed with Manfred Schlosser at Lausanne; his research on alkali–organic and organophosphorus compounds is well-known. Our relation goes back to several visits at the University of Lausanne, where I presented courses such as "Reactive Intermediates" and "Chemical Kinetics–the Crucial Test for Reaction Mechanisms". French is the principal language in Western Switzerland. Because my versatility in French was insufficient, I suggested German or English for my discourses. The audience's choice was English, although German is the major Swiss language.

In 1989–1991 I was a *professore a contratto* (visiting professor) at the University of Milan three times, each time giving a 15-hour course on "Orbital Symmetry in Organic Chemistry" or on "Chemical Kinetics". Franco Sannicolò, a former student of Raffaello Fusco, was my host; his research on hydrazo rearrangements, the Wallach reaction, and various aspects of cycloadditions offered material for heated mechanistic discussions.

In 1993 I was active at the University of Pavia in the same role, and Pierluigi Caramella was my host. He had worked with me as a postdoctoral fellow and later was introduced into theoretical chemistry by Ken Houk. Pavia, with its medieval center, is a temptation for strolling. It was a delight for Trudl and me to stay in Collegio Ghislieri, a lovely 17th century Baroque building.

On the German Education of Chemistry Students

A critical evaluation of the educational system is prompted by one's opportunity to compare. I looked into American education on my first trip in 1955, and my reaction at the time was: "How can two systems that differ so widely work equally well?" The contact deepened during visiting professorships; all in all, I spent more than 2.5 years in the United States.

The strength and backbone of the German approach are the laboratory courses, quite in contrast to the American system. Originally, the number of preparations, analyses, etc., was fixed, and the individual student determined the pace. In addition, the curriculum contained lecture courses; they were sometimes less effective because no tests were required and the temptation of absenteeism was great. Two oral examinations, *Vordiplom* and *Diplom*, mark the midpoint and end of the undergraduate experience.

That German system, more liberal than the American system, worked well for gifted and motivated students. When the West German *Wirtschaftswunder* ("economic miracle") brought affluence and the number of students soared, our educational system lost its edge. The number of years spent at the university increased in inverse proportion to individual accomplishments. Among the rationalizations I initiated in Munich in the 1960s to restrict this time were a final test of the introductory lecture course and a reorganization of the beginners' laboratory course in organic chemistry; features of this course are obligatory attendance, clearly defined experimental programs, accompanying lectures, and weekly tests. In these adaptations to the American system, with its resemblance to the high school approach, our Munich Institute was perhaps the forerunner. However, similar regulations are now ubiquitous throughout Germany.

The official curriculum prescribes eight semesters for undergraduate training plus one semester for the diploma thesis. With ineffective enforcement, 16 semesters are no rarity. The German average in 1990 was 13.2 semesters.[608]

The authors of the German Constitution after World War II introduced equal opportunity by guaranteeing free education for all. The German student pays no tuition, scholarships help the needy, and all laboratory expenses are paid by the university. Since 1982, scholarships are partially granted as loans; they place only minimal conditions on academic achievements. According to a German proverb, things that come free aren't worth anything. Personally, I am not happy with the system; I would prefer a generous support of those students who truly excel.

German universities still work within the classic academic year. Two semesters of undergraduate education cover 6.5 months, yet the average undergraduate today can hardly make proper use of the remaining 5.5 months for private study. I regard the quarter system practiced at many United States universities as superior.

Concerning the two decisive undergraduate oral exams, it is hardly conceivable that nine examiners in organic chemistry—the present status in Munich—would have similar rating standards. Each candidate is examined by only one out of ten. The participation of two examiners for each student would diminish the dilemma, but even this suggestion did not find a consensus. The student, with the right of proposing his or her own examiners, takes advantage of disparities. A written test would allow a more objective evaluation. Furthermore, a written test more closely resembles what is required of the professional chemist in research, development, and management. Thorough thinking is valued over quickness.

After the diploma exam, more than 90% of German chemistry students go on for graduate work. This is a chemistry-specific feature; the German chemical industry did not regard the diploma chemist (diploma introduced in 1939) as sufficiently educated. Specialization is supposed to begin in the graduate period. Lectures should accompany the experimental work for the diploma and Ph.D. thesis. Such graduate courses are offered, at least in Munich, but attendance is meager because of a complete lack of enforcement. I voiced my concern,[612] but the response was next to nil.

With the number of chemistry students rising, a slow but steady erosion of the requirements in practical courses and exams set in. In the 1960s a still-intact Ph.D. exam provided the graduate student with an incentive to read current literature and to attend lectures and colloquia. In the present version of the exam, however, the candidates report on their own thesis work. The questions of the examiners, who are chosen by the candidate, usually remain in that specific area. Moreover, inflation of grades is hardly an impetus for the graduate student to acquire knowledge. To my amazement, this degeneration of the Ph.D. exam in chemistry has seen parallel development at most German institutes.

According to 1992 statistics,[608] the average German chemistry student required 19.1 semesters (varying from 12 to 31 semesters) to reach the Ph.D. Thus, the new graduate leaves the university at an average age of 30 versus 26 in the United States; 4 years of elementary school and 9 years of high school plus military service precede the university education.

In my opinion the American system, free of the trammels of tradition, is better adapted to the modern student in an affluent society. However, when I discuss the topic with friends from the United States, I hear bitter complaints about American chemical education. The grass is always greener on the other side.

In Germany, supply and demand of professional chemists slipped out of balance in recent years. With their expanded capacity, the universities "overproduced" chemists. The number of beginners doubled from 1975 to 1990 but has been falling since then.[608] On the other hand, the pent-up demand of the chemical industry is now satisfied. Industry increasingly invests in foreign countries (one-third of research investments in 1991[608]), partly a result of the widespread phobia about chemistry in Germany. In 1990 German industry hired 47% of the fresh Ph.D. chemists; in the recession years 1991 and 1992 that figure shrank to 33% and 23%, respectively. In 1993, nearly 3700 Ph.D. chemists are seeking employment. Conceivably, German chemists might evolve into an (unprofitable) export article.

A feature of the German educational system that I am willing to defend is the *habilitation* (i.e., the procedure by which a promising young scientist enters an academic career). The candidate is granted up to 5 years to come to grips with an important research project, without any pressure to publish during that period. The research chosen should not be related to one's own doctoral or postdoctoral work.

Laboratory Equipment and Funding, Then and Now

In 1946–1947 I did the research for my *habilitation* in the temporary laboratory at Weilheim, Upper Bavaria (page 5). The makeshift facilities hardly resembled the luxury of the modern laboratory. This early experience and the need to improvise had a lifelong impact on me. My inclination to economize on research expenditures was certainly influenced by Heinrich Wieland. Consequently, my research work has always been relatively inexpensive.

The predicament of the early postwar years left us no other choice than to construct thermostats and instruments for dielectric and polarographic measurements ourselves. In that respect, I am grateful to Helmut Walz (1924–1983), a student of G. Kortüm at Tübingen, who followed me to Munich and operated our physical organic laboratory from 1952 to 1956. No less am I indebted to Helmut Huber, who has been in charge of the institute's physical instruments since 1962. Although he is a lab technician, Huber is in no way second to people with a university degree. Quite a number of my publications were coauthored by Huber.

In the 1960s and later the institute bought commercial instruments in strict accord with demand and frequency of use. I have retained a Spartan streak, in that I am still bothered by expensive equipment that is not used to the full extent. On "sightseeing tours" I was occasionally shown expensive apparatus under dusty plastic covers, guarded in locked rooms. Do such acquisitions serve the prestige of an institute and its professor? Personally, I vote for the primacy of the creative idea over the apparatus.

Several times my studies required flash photolysis. Instead of buying the equipment, I preferred to cooperate with specialists. Ernst Otto Fischer, associate professor at the adjacent Institute of Inorganic Chemistry in the early 1960s, received the first NMR instrument in our university and kindly allowed my co-workers to run spectra. Two years later Fischer suggested that I get my own spectrometer because my students used the Varian A60 more than his did. I consented and, of course, the new machine subsequently served the needs of all research groups of the institute.

NMR spectroscopy has revolutionized the daily life of the organic chemist. In a 1979 letter my friend Fritz Kröhnke (1903–1981), professor emeritus at Giessen, decried the changed working habits: "Remarkable and new the approach of the graduate students; first some intelligent considerations, then they pour the reactants into an NMR tube and hurry to the instrument. No perception for material phenomena, no more proficiency in the classic art of experimentation, nothing but experience in spectroscopy." Yet in quantitative analysis, detection of intermediates at low temperature, time-dependent unfolding of complex reactions, and line-shape analysis of rate constants the unique role of NMR spectroscopy can hardly be overestimated. For nearly 20 years 60-MHz spectroscopy became part of our lives. Now instruments with up to 600 MHz and two-dimensional techniques open up new territories.

Today no organic chemistry laboratory is competitive without appropriate NMR equipment. When cryomagnetic NMR spectrometers became available in the 1980s, the purchase of an instrument were no longer under my control. I sent out samples to benevolent colleagues to get spectra. Finally, in 1990 a Varian 400-MHz machine was set up in our laboratory. I am excited about the quality of the spectra, and I soon found quantitative ^{19}F NMR analysis of multicomponent mixtures indispensable for solving recent cycloaddition problems.

I am likewise enthusiastic about a new mass spectrometer that prints intensities with two decimals and makes high resolution easier than before. Since 1990 I routinely use ^{13}C and ^{34}S isotope peaks in assigning fragments.

Many of our Ph.D. students have teaching assistantships and help in the undergraduate laboratory courses. Besides the chemistry majors, about 350 medical students per semester take a laboratory course in chemistry in our institute; this enrollment increases the number of assistants needed. These assistantships pay very well today, but the remuneration was low in the 1950s and usually had to be split by two Ph.D. students. In the 1960s my co-workers told me—tongue in cheek—that the salary was sufficient to either buy a secondhand car or get married. A wife with her own job was obviously not as common then as it is today.

The institute budget takes care of the cost of energy and other operational expenses and, to a certain extent, covers glassware and chemicals. Most scientists with larger research groups have to apply for funds to the *Deutsche Forschungsgemeinschaft* (DFG, German Research Association) or to the Federal Ministry for Science and Technology. The DFG is an excellent organization, more or less run by scientists. However, special programs have tied up so much money that at present only 30% of the general applications for personnel and material can be served. Although the grant situation is very tough in Germany, there is a consolation. We do not have quite as much red tape as in the United States, where I shuddered to hear of applications submitted 24-fold.

On German Institutes, Past and Present

At the beginning of this century there was one professor of chemistry at each university. Several *Professors extraordinarius* (associate professors) and *Dozenten* (lecturers) shared teaching duties but were completely independent in their research.

After World War II, one full professorship each was assigned to inorganic, organic, physical, and biochemistry. Correspondingly, the chemical institute was split into independent units as to administration and budget for the various disciplines of chemistry. Most German universities still retain the institute system.

Traditionally, a young scientist obtained a Ph.D. and habilitated at the same institute. With the habilitation (page 236), the title *Dozent* (from the Latin *docere*, for teaching) was acquired. The *Dozent* continued independent research until offered an associate professorship by another university. This position used to be transitory, on the way to full professorship at a third university. Thus, the professional ladder was not climbed step by step at the home university. The so-called *Hausberufung* ("in-house appointment") had a touch of an "odor" and was finally forbidden in the new university constitutions established in the late 1970s.

The *ordinarius* (full professor) becomes emeritus at the age of 68. This step ends active teaching, but the professor emeritus retains an office and a research laboratory. Unfortunately, the "emeritus right" is no longer part of contracts issued since 1977. In the future full professors will be pensioned at the age of 65 like other state officials.

In 1962 politicians declared a *Bildungsnotstand* (educational crisis), and soon crowds of students flooded the universities. Within 15 years the number of German universities has doubled. The traditional schools were enlarged by doubling or tripling all professorial ranks. The number of German students reached 1.9 million in 1993 (with 85 million inhabitants in reunified Germany). In 1990, 34% of the youngsters in the appropriate age group enrolled at the universities. In contrast, hundreds of thousands of apprenticeships in the crafts and trades remained vacant. Germany still has brilliant students, but their absolute number has not increased.

The danger of such hasty development is obvious: quantity at the expense of quality. In the wake of the "student revolution" (1969–1973), several universities indulged in excesses like third parity (i.e., equal representation of professors, teaching assistants, and students on all decision-making boards). Fortunately, third parity was revoked by the German Constitutional Court, and the "democratization of the university" came to a halt. During that period I learned how much the efficiency of an educational system depends on its organizational structure.

American friends commented that German university research may have lost its clout with the transition to the multiprofessor system. Admittedly, Hans Fischer's (1881–1945) monumental work on structure and synthesis of porphyrins required the concentrated effort of the whole institute at the *Technische Hochschule* (Technical University) in Munich. On the other hand, diversified research activity foments fruitful discussion and stimulation. The multiprofessor system will work well if the participants share common values and are equally dedicated to the success of the institution.

The excellence of American higher education rests on the recognized and accepted quality differentiation of universities. Some elite universities are the pride of England, but elitist thinking is taboo at German schools of higher education. Wide sections of the population, especially the youth, are fascinated by the idea of égalité.

I often notice that young scientists have an unlimited faith in academic institutions. Their permanence is not taken for granted by my generation. The quality of academic institutions, no less than other values we cherish, can be preserved only through constant effort.

My description of the German institute system may be supplemented by brief notes on the Munich Institute and my own role. My associate professorship at Tübingen was a transitory position. The return to Munich was not considered a *Hausberufung* because I was offered full professorships by three universities in 1951–1952. When I arrived in Munich in 1952, the staff of the Institute of Organic Chemistry consisted of one *ordinarius* (full professor), one *extraordinarius* (associate professor), and five *Dozenten* (lecturers) with tenure.

The number of *Dozenten* was high. During the war the merry-go-round of professorial placements creaked and came to a halt. I was the youngest staff member in 1952; "Cassandra calls" spelling trouble did not materialize. The cooperation in rebuilding the educational system and in curricular reform was harmonious. I am grateful to my early colleagues: Hans Behringer, Alfred Bertho, Elisabeth Dane, Rudolf Hüttel, and Friedrich Klages.

Good secretarial help saves time, trouble, and strength. I express my thanks to Christine Rieger and Maria Kluge, who assisted me with great efficiency during my nearly 40 years in Munich. When an institute is regarded as well run, it may be largely the accomplishment of a single competent secretary.

During my residence time, the associate professorship was filled by Erich Schmidt (1923–1956), Hermann Stetter (1955–1960), Siegfried Hünig (1960–1961), Klaus Hafner (1962–1964), and Rudolf Gompper (1965–1968). The latter was promoted to *ordinarius* in 1968 when the professorial ranks swelled in the wake of the *Bildungsnotstand* (educational crisis).

The staff of the Institute of Organic Chemistry at Munich presently consists of three full professors (C4 positions) and six associate (C3) and assistant professors (C2), all independent in their research. A

Maria Kluge (left) and Christine Rieger deserve the flowers. Mrs. Rieger (1905–1989) served the institute for nearly 50 years and was secretary to Heinrich Wieland. She passed on to me my predecessor's advice to handle unagreeable matters without delay, but I did not always comply. The photograph was taken on June 13, 1970, my fiftieth birthday. The crowning touch to secretarial empathy: Christine Rieger went so far as to have her birthday on the same day of the year. In 1966 Mrs. Kluge succeeded her as secretary; her activity is a blessing for professor and students.

Faculty of Chemistry and Pharmacy deals with curricular questions, exam regulations, and academic personnel.

I became professor emeritus in 1988 after 36 years "on duty". I had no difficulty in detaching from institute affairs. I now cooperate with a small research group including a technician, and a part-time secretary helps me with the correspondence and the typing of my publications.

Settling my succession was not a quick operation, regrettably, despite the generous help offered by the Bavarian minister of education. In 1991 the Chair was finally taken over by Wolfgang Steglich of Bonn, who is widely known for his excellent studies on pigments and biologically active compounds of mushrooms. In addition, his astounding range of research interests includes preparative and mechanistic prob-

Would Adolf von Baeyer (1835–1917; bronze by H. Hahn, 1915) approve of his successors? Wolfgang Steglich (left) and I in the courtyard of the Munich Institute in 1992.

lems as well as synthesis. He promises a good future to the institute. Thus, Munich's tradition in natural product chemistry will be continued after an interlude of physical organic chemistry (1952–1988).

The new *ordinarius* receives a nonrecurring budget for his own research equipment, as well as for instruments serving the entire institute. In 1991 this "dowry" was more than 100 times higher than in 1952. *Tempora mutantur* (Times change), and scientific research has not become less expensive.

Has Chemistry Reached the Postmechanistic Era?

We are living in a fast-moving age. Yesterday's great fashions are oldies today. Special events in political, social, and cultural life serve as markers in the flow of time and divide the continuum into handy sections. The prefixes "pre" and "post" indicate the relation to such milestones. History is preceded by prehistory, and in the recent post-cold-war time enthusiastic writers hastily predicted the end of history, posthistory. Unpleasant events have squelched that dream in the meantime.

Science is not free of markers and turning points. Fashions come and go in both ladies' apparel and scientific research. Regrettably or not, reaction mechanisms are no longer the height of fashion. The general interest has swung to synthetic methods and the synthesis of natural products.

A recent episode is symptomatic. The American Chemical Society routinely monitors its publications to ensure that they fully serve the current information needs. In 1990 I was one of the test persons evaluating the *Journal of the American Chemical Society*. In the question-naire I had to check one box out of 17 for my primary subdiscipline.

The thesis work of Johann Mulzer (Ph.D. 1974) dealt with stereochemistry and mechanism of the rearrangement of styrylcyclopropanes (not described in this volume). Mulzer habilitated at Munich in 1980 with a study of aldol reactions of ester enolates. Since 1984 a professor at the Free University of Berlin, he ventured into grand-scale synthesis of natural products and recently completed a new route to erythronolide B.

Fields like bioorganic, computational, and theoretical chemistry were listed, but where would reaction mechanisms and physical organic chemistry fit in? I felt somewhat pushed over the edge. Apparently the "postmechanistic era" catchword is no joke. Sometimes the heretical notion is heard that reaction mechanisms are sufficiently known and can never be proven anyway.

A fateful consequence is the problem of financing research. Wherever I came, I heard resounding complaints about insufficient funding of research on reaction mechanisms. Several universities searched in vain for a competent scientist teaching reaction mechanisms.

The monolith of R. B. Woodward's ingenious syntheses of complex natural products redirected the flow of research for a generation. When I asked Bob Woodward in 1961 why he had wanted to synthesize chlorophyll, the roguish answer was "because nobody else could do it." Today asymmetric synthesis is adroitly used, and the synthetic goal sparked the elaboration of many a preparative method.

Ernst Biekert (left), president of the Gesellschaft Deutscher Chemiker, *presenting the Otto Hahn Prize for Chemistry and Physics to me in Berlin in 1979. Former recipients were H. Wieland (1955), L. Meitner (1955), H. Meerwein (1959), M. Eigen (1962), E. Hückel (1965), G. Wittig (1967), and F. Hund (1974).*

Alpinists have chosen certain mountains like Eiger Northside for measuring their skills. Similarly, synthetic chemists are vying for specific molecules harboring a top degree of complexity. Commercial interest is sometimes claimed in grant applications. Who wants to transfer the total syntheses of erythromycine or maytansine to industrial scale? It is hard to compete with fermentation; microorganisms produce less waste and do not need free weekends.

The elegant and innovative synthesis will remain a domain of the masters. In my opinion, the mere stringing of known reaction steps for building complex natural products is not the most rational use of time and funds, and I hope the fashion will soon swing to more rewarding areas of research.

Reaction mechanisms deserve a renaissance. Ludwig Boltzmann's catch phrase, "Nothing is more practical than a good theoretical concept", still holds true. The richer the knowledge about a reaction pathway, the higher the chances of superior control, both in the laboratory procedure and on the industrial scale. The factors determining reactivity sequences of nucleophiles, radicals, electrophiles, cycloaddition partners, etc., are insufficiently understood. Many sections of our science will profit from a better interpretation of selectivity phenomena.

Reaction mechanisms comprise a bridge-building subdiscipline that provides unifying concepts. Enzyme catalysis and much of the present bioorganic studies deal with mechanisms of organic molecules. The boundary between chemistry and life science is losing its distinction. The unveiling of biogenetic pathways will continue to fascinate. Enzyme activity, a model of catalytic processes, makes us humble. We should not give up on the far-away goal, understanding interactions with the protein surface and, perhaps, utilizing it in artificial enzymes.

> *Scientists have searched for a* perpetuum mobile; *they have found it: it is science itself.*
>
> Victor Hugo (1802–1885)

The French poet used this appealing image 120 years ago to state that science is an everlasting mission.

Acknowledgment

"To describe the development of specific areas of organic chemistry, to highlight the crucial discoveries, and to examine the impact they have had on the continuing development in the field" were among the goals the series editor had outlined in July 1986. Each of 24 authors was to contribute 25–40 printed pages toward a single book. My frantic endeavors to keep the text short led to an indigestibly compact report from which the personal touch had vanished. When I submitted the first sections of my article to the editor, he returned it marked with numerous requests for more scientific information and for additional personal comments: His "do not worry about space" (April 1987) resolved my dilemma.

Jeff Seeman's literary enterprise was without precedent; necessary changes in editorial policy are well excusable. I wish to express my sincere thanks to him for his continuously polite, tactful, and encouraging guidance. The original warning against an "overload of biographical introspection" was soon replaced by the plea for more and more autobiographical data. I concentrated these supplements in a concluding chapter that kept growing over the years. Two biographical chapters now frame 10 largely technical chapters. The technical chapters offer more detail than do the previously published volumes of this series; I hope the reader will not disapprove of this deviation.

My limited command of English, a language I learned in my late twenties, did not meet my own demands. I am deeply grateful to my daughters for their generous help: Helga Huisgen of Munich corrected most of the text stylistically, and Professor Birge Zimmermann-Huisgen of Santa Barbara smoothed out much of the two biographical sections. Not less indebted am I to Colleen Stamm, Janet Dodd, Robin Giroux, and Margaret Brown of the American Chemical Society Books Department, for further stylistic corrections.

How far should extra scientific interests enter a scientific autobiography? Again I owe thanks to Jeff Seeman who encouraged me to voice my thoughts on many subjects not directly chemical, such as 'women in science' and art. He even supported the incorporation of a drawing and a poem by my friend and late brother Klaus.

Finally, Maria Kluge deserves thanks for typing the first draft in 1987 and the supplemental version in 1989; corrections and inserts in 1992 demanded more of her patience. Doris Reindl drew the formula blocks.

References

1. Walden, P. *Geschichte der Organischen Chemie seit 1880;* Julius Springer: Berlin, Germany, 1941; 946 pp.
2. Fischer, F. G. "Heinrich Wieland, 1877–1957" (Obituary); *Jahrbuch Bayer. Akad. Wissenschaften* **1959**, 160–169.
3. *Survey:* Huisgen, R. *Angew. Chem.* **1950**, *62*, 527–534.
4. (a) Huisgen, R. "The Wieland Memorial Lecture"; *Proc. Chem. Soc.* **1958**, 210–219; (b) *Review:* Franke, W. *Naturwissenschaften* **1942**, *30*, 342–351; (c) Braun, T. *Angew. Chem. Int. Ed. Engl.* **1992**, *31*, 473.
5. Huisgen, R. *Justus Liebigs Ann. Chem.* **1948**, *559*, 101–152.
6. Huisgen, R. *Justus Liebigs Ann. Chem.* **1949**, *564*, 16–32.
7. Wieland, H. *Justus Liebigs Ann. Chem.* **1934**, *514*, 145–181.
8. Wieland, H.; Ploetz, Th.; Indest, H. *Justus Liebigs Ann. Chem.* **1937**, *532*, 166–190.
9. Hey, D. H.; Waters, W. A. *Chem. Rev.* **1937**, *21*, 169–208.
10. Grieve, W.; Hey, D. H. *J. Chem. Soc.* **1934**, 1797–1806.
11. Bamberger, E. *Ber. Dtsch. Chem. Ges.* **1894**, *27*, 914–917.
12. Hey, D. H.; Waters, W. A. *J. Chem. Soc.* **1948**, 882–885.
13. Huisgen, R.; Horeld, G. *Justus Liebigs Ann. Chem.* **1949**, *562*, 137–162.
14. Huisgen, R. *Justus Liebigs Ann. Chem.* **1951**, *574*, 184–201.
15. Huisgen, R.; Nakaten, H. *Justus Liebigs Ann. Chem.* **1954**, *586*, 70–83.
16. Grashey, R.; Huisgen, R. *Chem. Ber.* **1959**, *92*, 2641–2545.
17. Huisgen, R.; Grashey, R. *Justus Liebigs Ann. Chem.* **1957**, *607*, 46–59.
18. Hey, D. H.; Stirling, C. J. M.; Williams, G. H. *J. Chem. Soc.* **1955**, 3963–3969.
19. Huisgen, R.; Jakob, F.; Grashey, R. *Chem. Ber.* **1959**, *92*, 2206–2210.
20. Barben, I. K.; Suschitzky, H. *J. Chem. Soc.* **1960**, 2735–2739.
21. Rüchardt, C.; Tan, C. C.; Freudenberg, B. *Tetrahedron Lett.* **1968**, 4019–4022.
22. Rüchardt, C.; Freudenberg, B. *Tetrahedron Lett.* **1964**, 3623–3628.
23. Chalfont, G. R.; Perkins, M. J. *J. Am. Chem. Soc.* **1967**, *89*, 3054–3055.
24. Cadogan, J. I. G.; Paton, R. M.; Thomson, C. *J. Chem. Soc. B* **1971**, 583–595.
25. Huisgen, R.; Krause, L. *Justus Liebigs Ann. Chem.* **1951**, *574*, 157–171.

26. Huisgen, R.; Reimlinger, H. *Justus Liebigs Ann. Chem.* **1956**, *599*, 161–182.
27. Huisgen, R.; Reinertshofer, J. *Justus Liebigs Ann. Chem.* **1952**, *575*, 197–216.
28. Huisgen, R. *Justus Liebigs Ann. Chem.* **1951**, *574*, 171–184.
29. Huisgen, R. *Justus Liebigs Ann. Chem.* **1951**, *573*, 163–181.
30. von Pechmann, H. *Ber. Dtsch. Chem. Ges.* **1894**, *27*, 1888–1891.
31. Regitz, M.; Maas, G. *Diazo Compounds, Properties and Synthesis*; Academic: Orlando, FL, 1986; pp 301–322.
32. Huisgen, R. In *Methoden der Organischen Chemie (Houben–Weyl)*, 4th ed.; Müller, E., Ed.; G. Thieme: Stuttgart, Germany, 1955; Vol. 3/1, pp 101–162.
33. Zollinger, H. *Azo and Diazo Chemistry*; Interscience: New York, 1961; Chapter 10.
34. Huisgen, R.; Nakaten, H. *Justus Liebigs Ann. Chem.* **1954**, *586*, 84–109.
35. Wittwer, C.; Zollinger, H. *Helv. Chim. Acta* **1954**, *37*, 1954–1968.
36. Jacobson, P.; Huber, L. *Ber. Dtsch. Chem. Ges.* **1908**, *41*, 660–671.
37. Huisgen, R.; Bast, K. *Org. Synth.* **1962**, *42*, 69–72; see Note 12.
38. Huisgen, R.; Nakaten, H. *Justus Liebigs Ann. Chem.* **1951**, *573*, 181–195.
39. Tröndlin, F.; Werner, R.; Rüchardt, C. *Chem. Ber.* **1978**, *111*, 367–378.
40. *Review:* Cadogan, J. I. G. *Acc. Chem. Res.* **1971**, *4*, 186–192.
41. Cadogan, J. I. G.; Murray, C. D.; Sharp, J. T. *J. Chem. Soc. Perkin Trans. 2* **1976**, 583–587.
42. Rüchardt, C.; Tan, C. C. *Angew. Chem. Int. Ed. Engl.* **1970**, *9*, 522–523.
43. Hassmann, V.; Rüchardt, C.; Tan, C. C. *Tetrahedron Lett.* **1971**, 3885–3887.
44. Huisgen, R.; Reimlinger, H. *Justus Liebigs Ann. Chem.* **1956**, *599*, 183–202.
45. Stangl, H. Ph.D. Thesis, University of Munich, 1959.
46. Reimlinger, H. Ph.D. Thesis, University of Munich, 1953.
47. Huisgen, R.; Rüchardt, C. *Justus Liebigs Ann. Chem.* **1956**, *601*, 1–21.
48. Burr, J. G.; Ciereszko, L. S. *J. Am. Chem. Soc.* **1952**, *74*, 5426–5433.
49. Foster, M. S.; Beauchamp, J. L. *J. Am. Chem. Soc.* **1972**, *94*, 2425–2431.
50. Ford, G. P. *J. Am. Chem. Soc.* **1986**, *108*, 5104–5108.
51. Semenov, D.; Shih, C. H.; Young, W. G. *J. Am. Chem. Soc.* **1958**, *80*, 5472–5475.
52. Kirmse, W.; Voigt, G. *J. Am. Chem. Soc.* **1974**, *96*, 7598–7599.
53. Huisgen, R.; Rüchardt, C. *Justus Liebigs Ann. Chem.* **1956**, *601*, 21–39.
54. Streitwieser, A.; Schaeffer, W. D. *J. Am. Chem. Soc.* **1957**, *79*, 2888–2893.
55. McGarrity, J. F.; Cox, D. P. *J. Am. Chem. Soc.* **1983**, *105*, 3961–3966.
56. *Review:* Kirmse, W. *Angew. Chem. Int. Ed. Engl.* **1976**, *15*, 251–261.
57. Feldmann, G.; Kirmse, W. *Angew. Chem. Int. Ed. Engl.* **1987**, *26*, 571–572.
58. (a) Curtius, Th. *Ber. Dtsch. Chem. Ges.* **1883**, *16*, 2230–2231; (b) *Review:* Huisgen, R. *Angew. Chem.* **1955**, *67*, 439–463.
59. Huisgen, R.; Koch, H. J. *Justus Liebigs Ann. Chem.* **1955**, *591*, 200–231.
60. Clusius, K.; Hürzeler, H.; Huisgen, R.; Koch, H. J. *Naturwissenschaften* **1954**, *41*, 213.
61. Huisgen, R.; Fleischmann, R. *Justus Liebigs Ann. Chem.* **1959**, *623*, 47–68.
62. Meyer, K. H. *Ber. Dtsch. Chem. Ges.* **1919**, *52*, 1468–1476.
63. Carlson, B. A.; Sheppard, W. A.; Webster, O. W. *J. Am. Chem. Soc.* **1975**, *97*, 5291–5293.
64. Bronberger, F.; Huisgen, R. *Tetrahedron Lett.* **1984**, *25*, 65–68.

65. Huisgen, R.; Bronberger, F. *Tetrahedron Lett.* **1984**, *25*, 61–64.
66. Stoll, M.; Stoll-Comte, G. *Helv. Chim. Acta* **1930**, *13*, 1185–1200.
67. *Review:* Prelog, V. *J. Chem. Soc.* **1950**, 420–428.
68. Traetteberg, M. *Acta Chem. Scand.* **1975**, B29, 29.
69. Fittig, R. *Justus Liebigs Ann. Chem.* **1881**, *208*, 111–121.
70. Wislicenus, W. *Justus Liebigs Ann. Chem.* **1886**, *233*, 101–116.
71. Huisgen, R.; Ott, H. *Tetrahedron* **1959**, *6*, 253–267.
72. O'Gorman, J. M.; Shand, W.; Schomaker, V. *J. Am. Chem. Soc.* **1950**, *72*, 4222–4228.
73. (a) Blom, C. E.; Günthard, H. H. *Chem. Phys. Lett.* **1981**, *84*, 267–271;
    (b) Wiberg, K. B.; Laidig, K. E. *J. Am. Chem. Soc.* **1987**, *109*, 5935–5943;
    (c) Wiberg, K. B.; Waldron, R. F. *J. Am. Chem. Soc.* **1991**, *113*, 7697–7705;
    (d) Brown, J. M.; Conn, A. D.; Pilcher, G.; Leitao, M. L. P.; Yang, M.-Y. *J. Chem. Soc. Chem. Commun.* **1989**, 1817–1819.
74. Stelter, H. Ph.D. Thesis, University of Munich, 1965.
75. Pauling, L.; Corey, R. B.; Branson, H. R. *Proc. Natl. Acad. Sci. USA* **1951**, *37*, 205–234, 235–240.
76. *Discussion of* cis-*configuration:* Pauling, L.; Corey, R. B. *Proc. Natl. Acad. Sci. USA* **1952**, *38*, 86–93.
77. Worsham, J. E.; Hobbs, M. E. *J. Am. Chem. Soc.* **1954**, *76*, 206–208.
78. Kotera, A.; Shibata, S.; Sone, K. *J. Am. Chem. Soc.* **1955**, *77*, 6183–6186.
79. Huisgen, R.; Walz, H. *Chem. Ber.* **1956**, *89*, 2616–2629.
80. Günthard, H. H.; Gäumann, T. *Helv. Chim. Acta* **1951**, *34*, 39–46.
81. Schiedt, U. *Angew. Chem.* **1954**, *66*, 609–610.
82. Huisgen, R.; Brade, R.; Walz, H.; Glogger, I. *Chem. Ber.* **1957**, *90*, 1437–1447.
83. *Review:* Hallam, H. E.; Jones, C. M. *J. Mol. Struct.* **1970**, *5*, 1–19.
84. Dunitz, J. D.; Winkler, F. K. *Acta Crystallogr. Sec. B Struct. Crystallogr. Cryst. Chem.* **1975**, *31*, 251–263.
85. Ziegler, K.; Eberle, H.; Ohlinger, H. *Justus Liebigs Ann. Chem.* **1933**, *504*, 94–130.
86. Huisgen, R.; Rapp, W.; Ugi, I.; Walz, H.; Glogger, I. *Justus Liebigs Ann. Chem.* **1954**, *586*, 52–69.
87. Huisgen, R.; Rapp, W. *Chem. Ber.* **1952**, *85*, 826–835.
88. Huisgen, R.; Ugi, I.; Rauenbusch, E.; Vossius, V.; Oertel, H. *Chem. Ber.* **1957**, *90*, 1946–1958.
89. Huisgen, R.; Ugi, I. *Chem. Ber.* **1960**, *93*, 2693–2704.
90. Huisgen, R.; Vossius, V. *Monatsh. Chem.* **1957**, *88*, 517–540.
91. Huisgen, R.; Rietz, U. *Chem. Ber.* **1957**, *90*, 2768–2784.
92. Huisgen, R.; Rietz, U. *Tetrahedron* **1958**, *2*, 271–288.
93. Trescher, V. Ph.D. Thesis, University of Munich, 1956.
94. Oertel, H. Ph.D. Thesis, University of Munich, 1959.
95. *Review:* Huisgen, R. *Angew. Chem.* **1957**, *69*, 341–359.
96. Gol'dfarb, Y. L.; Taits, S. Z.; Belen'kii, L. I. *Izv. Akad. Nauk. SSR Ser. Khim.* **1963**, 1451–1460; *Chem. Abstr.* **1963**, *59*, 15243 f.
97. Huisgen, R.; Rapp, W.; Ugi, I.; Walz, H.; Mergenthaler, E. *Justus Liebigs Ann. Chem.* **1954**, *586*, 1–29.
98. Huisgen, R.; Ugi, I.; Brade, H.; Rauenbusch, E. *Justus Liebigs Ann. Chem.* **1954**, *586*, 30–51.

99. Brown, H. C.; Borkowski, M. J. *Am. Chem. Soc.* **1952,** *74,* 1894–1902.
100. Huisgen, R.; Ugi, I. *Chem. Ber.* **1960,** *93,* 2693–2704.
101. Cram, D. J. *J. Am. Chem. Soc.* **1949,** *71,* 3863–3870, 3875–3883.
102. Winstein, S.; Brown, M.; Schreiber, K. C.; Schlesinger, A. H. *J. Am. Chem. Soc.* **1952,** *74,* 1140–1147.
103. Huisgen, R.; Rauenbusch, E.; Seidl, G.; Wimmer, I. *Justus Liebigs Ann. Chem.* **1964,** *671,* 41–57.
104. Seidl, G.; Huisgen, R. *Chem. Ber.* **1963,** *96,* 2730–2739.
105. Huisgen, R.; Seidl, G. *Chem. Ber.* **1963,** *96,* 2740–2749.
106. Huisgen, R.; Seidl, G. (a) *Angew. Chem.* **1957,** *69,* 390–391; (b) *Tetrahedron* **1964,** *20,* 231–241.
107. Baeyer, A.; Villiger, V. *Ber. Dtsch. Chem. Ges.* **1902,** *35,* 1189–1201.
108. Meerwein, H.; Van Emster, K. *Ber. Dtsch. Chem. Ges.* **1922,** *55,* 2500–2528.
109. Meisenheimer, J. *Ber. Dtsch. Chem. Ges.* **1921,** *54,* 3206–3213.
110. Meisenheimer, J.; Zimmermann, P.; v. Kummer, U. *Justus Liebigs Ann. Chem.* **1925,** *446,* 205–228.
111. *Review:* Nair, V. In *Small-Ring Heterocycles;* Hassner, A., Ed.; The Chemistry of Heterocyclic Compounds Series, Vol. 47, Part 1; Wiley: New York, 1983; pp 215–332.
112. Neber, P. W.; Burgard, A. *Justus Liebigs Ann. Chem.* **1932,** *493,* 281–294.
113. Chapman, A. W.; Howis, C. C. *J. Chem. Soc.* **1933,** 806–811.
114. Huisgen, R.; Witte, J.; Walz, H.; Jira, W. *Justus Liebigs Ann. Chem.* **1957,** *604,* 191–202.
115. Pearson, D. E.; Baxter, J. F.; Martin, J. C. *J. Org. Chem.* **1952,** *17,* 1511–1518.
116. Pearson, D. E.; Ball, F. *J. Org. Chem.* **1949,** *14,* 118–131.
117. Huisgen, R.; Witte, J.; Ugi, I. *Chem. Ber.* **1957,** *90,* 1844–1849.
118. Curtius, Th.; Heidenreich, K. *Ber. Dtsch. Chem. Ges.* **1894,** *27,* 773–774.
119. Stollé, R.; Leffler, K. *Ber. Dtsch. Chem. Ges.* **1924,** *57,* 1061–1063.
120. Diels, O.; Blom, J. H.; Koll, W. *Justus Liebigs Ann. Chem.* **1925,** *443,* 242–262.
121. Diels, O.; Alder, K. *Justus Liebigs Ann. Chem.* **1928,** *460,* 98–122.
122. Kharasch, M. S.; White, P. C.; Mayo, F. R. *J. Org. Chem.* **1938,** *3,* 33–47.
123. Huisgen, R.; Jakob, F.; Siegel, W.; Cadus, A. *Justus Liebigs Ann. Chem.* **1954,** *590,* 1–36.
124. Huisgen, R.; Jakob, F. *Justus Liebigs Ann. Chem.* **1954,** *590,* 37–54.
125. Alder, K.; Noble, T. *Ber. Dtsch. Chem. Ges.* **1943,** *76,* 54–57.
126. Diels, O. *Justus Liebigs Ann. Chem.* **1922,** *429,* 1–55.
127. Kenner, G. W.; Stedman, R. J. *J. Chem. Soc.* **1952,** 2089–2094.
128. Alder, K.; Pascher, F.; Schmitz, A. *Ber. Dtsch. Chem. Ges.* **1943,** *76,* 27–53.
129. Alder, K.; Söll, H.; Söll, H. *Justus Liebigs Ann. Chem.* **1949,** *565,* 73–99.
130. *Review:* Hoffmann, H. M. R. *Angew. Chem. Int. Ed. Engl.* **1969,** *8,* 556–577.
131. Arnold, R. T.; Dowdall, J. F. *J. Am. Chem. Soc.* **1948,** *70,* 2590–2591.
132. Huisgen. R.; Pohl, H. *Chem. Ber.* **1960,** *93,* 527–540.
133. Thaler, W. A.; Franzus, B. *J. Org. Chem.* **1964,** *29,* 2226–2235.
134. Franzus, B. *J. Org. Chem.* **1963,** *28,* 2954–2960.
135. Askani, R. *Chem. Ber.* **1965,** *98,* 2551–2555.
136. Schenck, G. O.; Kopp, H. R.; Kim, B.; Koerner von Gustorf, E. *Z. Naturforsch.* **1965,** *20b,* 637–639.
137. Alder, K.; Niklas, H. *Justus Liebigs Ann. Chem.* **1954,** *585,* 97–114.

138. Koerner von Gustorf, E.; White, D. V.; Kim, B.; Hess, D.; Leitich, J. *J. Org. Chem.* **1970**, *35*, 1155–1165.

139. Seymour, C. A.; Greene, F. D. *J. Am. Chem. Soc.* **1980**, *102*, 6384–6385.

140. Nelsen, S. F.; Kapp, D. L. *J. Am. Chem. Soc.* **1985**, *107*, 5548–5549.

141. Müller, E. *Ber. Dtsch. Chem. Ges.* **1914**, *47*, 3001–3023.

142. Breslow, R.; Yaroslavsky, C.; Yaroslavsky, S. *Chem. Ind. (London)* **1961**, 1961.

143. Fahr, E.; Königsdorfer, K.; Scheckenbach, F. *Justus Liebigs Ann. Chem.* **1965**, *690*, 138–146.

144. Fahr E.; Döppert, K.; Scheckenbach, F. *Justus Liebigs Ann. Chem.* **1966**, *696*, 136–144.

145. Bettinetti, G. F.; Grünanger, P. *Tetrahedron Lett.* **1965**, 2553–2557.

146. Bettinetti, G. F.; Capretti, L. *Gazz. Chim. Ital.* **1965**, *95*, 33–42.

147. Huisgen, R.; Fleischmann. R.; Eckell, A. *Tetrahedron Lett.* **1960**, No. 12, 1–4.

148. Huisgen, R.; Fleischmann, R.; Eckell, A. *Chem. Ber.* **1977**, *110*, 500–513.

149. Brandl, F.; Hoppe, W. *Z. Kristallogr.* **1967**, *125*, 80–91.

150. Huisgen, R.; Eckell, A. *Tetrahedron Lett.* **1960**, No. 12, 5–8.

151. Huisgen, R.; Fleischmann, R.; Eckell, A. *Chem. Ber.* **1977**, *110*, 514–521.

152. Flippen-Andersen, J. L.; Karle, I.; Huisgen, R.; Reissig, H. U. *Angew. Chem. Int. Ed. Engl.* **1980**, *19*, 906–907.

153. Schmitz, E. *Chem. Ber.* **1958**, *91*, 1495–1503.

154. Huisgen, R.; Grashey, R.; Laur, P.; Leitermann, H. *Angew. Chem.* **1960**, *72*, 416–417.

155. Schiffer, R. Ph.D. Thesis, University of Munich, 1966.

156. *Review:* Grashey, R. "Azomethine Imines"; In *1,3-Dipolar Cycloaddition Chemistry*; Padwa, A., Ed.; Wiley: New York, 1984; pp 733–817.

157. Eckell, A.; George, M. V.; Huisgen, R.; Kende, A. S. *Chem. Ber.* **1977**, *110*, 578–595.

158. Huisgen, R.; Eckell, A. *Chem. Ber.* **1977**, *110*, 522–539, 540–558.

159. *Review:* Huisgen, R. *Angew. Chem. Int. Ed. Engl.* **1963**, *2*, 565–598.

160. Grashey, R.; Huisgen, R.; Leitermann, H. *Tetrahedron Lett.* **1960**, No. 12, 9–13.

161. Huisgen, R.; Grashey, R.; Hauck, H.; Seidl, H. *Chem. Ber.* **1968**, *101*, 2043–2055, 2548–2558.

162. Huisgen, R.; Hauck, H.; Grashey, R.; Seidl, H. *Chem. Ber.* **1968**, *101*, 2568–2584.

163. Sims, J.; Houk, K. N. *J. Am. Chem. Soc.* **1973**, *95*, 5798–5800.

164. Eckell, A.; Huisgen, R. *Chem. Ber.* **1977**, *10*, 571–577.

165. Grashey, R.; Adelsberger, K. *Angew. Chem.* **1962**, *74*, 292–293.

166. Grashey, R.; Leitermann, H.; Schmidt, R.; Adelsberger, K. *Angew. Chem.* **1962**, *74*, 491.

167. *Review:* Huisgen, R. *Wiss. Z. Karl-Marx Univ. Leipzig, Math.–Naturwiss. Reihe* **1983**, *32*, 395–406.

168. Karle, I. L.; Flippen-Anderson, J. L.; Huisgen, R. *Acta Crystallogr. Sect. C Cryst. Struct. Commun.* **1985**, *41*, 1095–1100.

169. Finke, J. Ph.D. Thesis, University of Munich, 1984.

170. Sun, K. K. Ph.D. Thesis, University of Munich, 1965.

171. Grashey, R.; Huisgen, R.; Sun, K. K.; Moriarty, R. M. *J. Org. Chem.* **1965**, *30*, 74–79.

172. Oppolzer, W. *Tetrahedron Lett.* **1970**, 2199–2204.
173. *Review:* Oppolzer, W. *Angew. Chem. Int. Ed. Engl.* **1977**, *16*, 10–24.
174. Huisgen, R.; Blaschke, H.; Brunn, E. *Tetrahedron Lett.* **1966**, 405–409.
175. Morrison, D. C. *J. Org. Chem.* **1958**, *23*, 1072.
176. Ginsburg, V. A.; Vasil'eva, M. N.; Dubov, S. S.; Yakubovich, Y. A. *J. Gen. Chem. USSR (Engl. Transl.)* **1960**, *30*, 2834–2841.
177. Cookson, R. C.; Locke, J. M. *J. Chem. Soc.* **1963**, 6062–6064.
178. Brunn, E. Ph.D. Thesis, University of Munich, 1968, pp 36–37.
179. Goncalves, H.; Dormoy, J. R.; Chapleur, Y.; Castro, B.; Fauduet, H.; Burgada, R. *Phosphorus and Sulfur* **1980**, *8*, 147–151.
180. Grochowski, E.; Hilton, B. D.; Kupper, R. J.; Michejda, C. J. *J. Am. Chem. Soc.* **1982**, *104*, 6876–6877.
181. Brunn, E.; Huisgen, R. *Angew. Chem. Int. Ed. Engl.* **1969**, *8*, 513–515.
182. Huisgen, R.; Brunn, E.; Gilardi, R.; Karle, I. *J. Am. Chem. Soc.* **1969**, *91*, 7766–7767.
183. Johnson, A. W.; Tebby, J. C. *J. Chem. Soc.* **1961**, 2126–2130.
184. Schmidpeter, A.; Zeiss, W. *Angew. Chem. Int. Ed. Engl.* **1971**, *10*, 396.
185. Rodionova, L. S.; Galishev, V. A.; Chistokletov, V. N.; Petrov, A. A. *J. Gen. Chem. USSR Engl. Transl.* **1975**, *45*, 1623.
186. *Review:* Mitsunobu, O. *Synthesis* **1981**, 1–28.
187. Mitsunobu, O.; Eguchi, M. *Bull. Chem. Soc. Jpn.* **1971**, *44*, 3427–3430.
188. *Review:* Mukaiyama, T. *Angew. Chem. Int. Ed. Engl.* **1976**, *15*, 94–102.
189. Limpricht, H. *Ber. Dtsch. Chem. Ges.* **1874**, *7*, 1349–1367.
190. *Review:* Sauer, J.; Huisgen, R. *Angew. Chem.* **1960**, *72*, 294–315.
191. *Review:* Bunnett, J. F.; Zahler, R. E. *Chem. Rev.* **1951**, *49*, 273–412.
192. Roberts, J. D.; Simmons, H. E.; Carlsmith, L. A.; Vaughan, C. W. *J. Am. Chem. Soc.* **1953**, *75*, 3290–3291.
193. (a) Huisgen, R.; Rist, H. *Naturwissenschaften* **1954**, *41*, 358–359; *Justus Liebigs Ann. Chem.* **1955**, *594*, 137–158; (b) Roberts, J. D. "The Right Place at the Right Time"; In *Profiles, Pathways, and Dreams*; Seeman, J. I., Ed.; American Chemical Society: Washington, DC, 1990; pp 106–119.
194. Wittig, G.; Pohmer, L. *Chem. Ber.* **1956**, *89*, 1334–1351.
195. Wittig, G.; Ludwig, R. *Angew. Chem.* **1956**, *68*, 40.
196. Stiles, M.; Miller, R. G. *J. Am. Chem. Soc.* **1960**, *82*, 3802.
197. Stiles, M.; Miller, R. G.; Burckhardt, U. *J. Am. Chem. Soc.* **1963**, *85*, 1792–1797.
198. Friedman, L.; Logullo, F. M. *J. Am. Chem. Soc.* **1963**, *85*, 1549.
199. Wittig, G.; Hoffmann, R. W. *Angew. Chem.* **1961**, *73*, 435–436; *Chem. Ber.* **1962**, *95*, 2718–2728.
200. Campbell, C. D.; Rees, C. W. *J. Chem. Soc. C* **1969**, 742–747.
201. Hoffmann, R. W. *Dehydrobenzene and Cycloalkynes*; Verlag Chemie: Weinheim, Germany, 1967.
202. *Review:* Gilman, H.; Morton, J. W. *Org. React.* **1954**, *8*, 258–304.
203. Wittig, G.; Pieper, G.; Fuhrmann, G. *Ber. Dtsch. Chem. Ges.* **1940**, *73*, 1193–1197.
204. Wittig, G. *Naturwissenschaften* **1942**, *30*, 696–703.
205. Wittig, G.; Fuhrmann, G. *Ber. Dtsch. Chem. Ges.* **1940**, *73*, 1197–1218.
206. Jenny, E. F.; Roberts, J. D. *Helv. Chim. Acta* **1955**, *38*, 1248–1254.

207. Bottini, A. T.; Roberts, J. D. *J. Am. Chem. Soc.* **1957**, *79*, 1458–1462.

208. Jung, D. Ph.D. Thesis, University of Munich, 1962.

209. Huisgen, R.; Sauer, J. *Chem. Ber.* **1958**, *91*, 1453–1460.

210. Huisgen, R.; Sauer, J.; Mack, W.; Ziegler, I. *Chem. Ber.* **1959**, *92*, 441–449.

211. Huisgen, R.; König, H. (a) *Angew. Chem.* **1957**, *69*, 268; (b) *Chem. Ber.* **1959**, *92*, 203–213.

212. Huisgen, R.; König, H.; Bleeker, N. *Chem. Ber.* **1959**, *92*, 424–429.

213. Huisgen, R.; König, H.; Lepley, A. R. *Chem. Ber.* **1960**, *93*, 1496–1506

214. Hrutfiord, B. F.; Bunnett, J. F. *J. Am. Chem. Soc.* **1958**, *80*, 2021–2022, 4745.

215. Bergstrom, F. W.; Wright, R. E.; Chandler, C.; Gilkey, W. A. *J. Org. Chem.* **1936**, *1*, 170–178.

216. Hall, G. A.; Piccolini, R.; Roberts, J. D. *J. Am. Chem. Soc.* **1955**, *77*, 4540–4543.

217. Roberts, J. D.; Semenov, D. A.; Simmons, H. E.; Carlsmith, L. A. *J. Am. Chem. Soc.* **1956**, *78*, 601–611.

218. Huisgen, R.; Sauer, J. *Chem. Ber.* **1959**, *92*, 192–202.

219. Huisgen, R.; Mack, W. *Chem. Ber.* **1960**, *93*, 332–340.

220. Wittig, G.; Hoffmann, R. W. *Chem. Ber.* **1962**, *95*, 2729–2734.

221. Huisgen, R.; Mack, W.; Herbig, K.; Ott, N.; Anneser, E. *Chem. Ber.* **1960**, *93*, 412–424.

222. Sunthankar, S. V.; Gilman, H. *J. Org. Chem.* **1951**, *16*, 8–16.

223. *Review:* Huisgen, R. *Angew. Chem. Int. Ed. Engl.* **1970**, *9*, 751–762.

224. Huisgen, R.; Mack, W.; Möbius, L. *Tetrahedron* **1960**, *9*, 29–39.

225. Huisgen, R.; Sauer, J.; Hauser, A. *Chem. Ber.* **1958**, *91*, 2366–2374.

226. Huisgen, R.; Knorr, R. *Tetrahedron Lett.* **1963**, 1017–1021.

227. Möbius, L. Ph.D. Thesis, University of Munich, 1961.

228. Montgomery, L. K.; Applegate, L. E. *J. Am. Chem. Soc.* **1967**, *89*, 5305–5307.

229. *Review:* Huisgen, R.; Sauer, J. *Angew. Chem.* **1960**, *72*, 91–108.

230. Mack, W.; Huisgen, R. *Chem. Ber.* **1960**, *93*, 608–613.

231. Huisgen, R.; Zirngibl, L. *Chem. Ber.* **1958**, *91*, 1438–1452, 2375–2382.

232. Hoffmann, R. W.; Vargas-Nunez, G. E.; Guhn, G.; Sieber, W. *Chem. Ber.* **1965**, *98*, 2074–2085.

233. Herbig, K. Ph.D. Thesis, University of Munich, 1961.

234. Kauffmann, Th.; Wirthwein, R. *Angew. Chem. Int. Ed. Engl.* **1971**, *10*, 20–33.

235. van der Plas, J. C.; Roeterdink, F. In *The Chemistry of Triple-Bonded Functional Groups*; Patai, S.; Rappoport, Z., Eds.; Wiley: New York, 1983; Supplement C, pp 421–511.

236. Reinecke, M. G. In *Reactive Intermediates*; Abramovitch, R. A., Ed.; Plenum: New York, 1981; Vol. 2.

237. Hoffmann, R.; Imamura, A.; Hehre, W. J. *J. Am. Chem. Soc.* **1968**, *90*, 1499–1509.

238. Washburn, W.; Zahler, R.; Chen, I. *J. Am. Chem. Soc.* **1978**, *100*, 5863–5874.

239. Zincke, Th.; Arzberger, H. *Justus Liebigs Ann. Chem.* **1888**, *249*, 350–372.

240. Hantzsch, A. *Ber. Dtsch. Chem. Ges.* **1903**, *36*, 2056–2058.

241. Dimroth, O.; de Montmollin, G. *Ber. Dtsch. Chem. Ges.* **1910**, *43*, 2904–2915.

242. Curtius, Th.; Darapsky, A.; Müller, E. *Ber. Dtsch. Chem. Ges.* **1915**, *48*, 1614–1634.

243. Huisgen, R.; Ugi, I. *Angew. Chem.* **1956**, *68*, 705–706.

244. Ugi, I.; Huisgen, R.; Clusius, K.; Vecchi, M. *Angew. Chem.* **1956,** *68,* 753–754.
245. Huisgen, R.; Ugi, I. *Chem. Ber.* **1957,** *90,* 2914–2927.
246. Ugi, I.; Perlinger, H.; Behringer, L. *Chem. Ber.* **1958,** *91,* 2324–2329.
247. *Review:* Tišler, M. *Synthesis* **1973,** 123–136.
248. Ugi, I.; Huisgen, R. *Chem. Ber.* **1958,** *91,* 531–537.
249. Wallis, J. D.; Dunitz, J. D. *J. Chem. Soc. Chem. Commun.* **1983,** 910–911.
250. *Review:* Huisgen, R. *Angew. Chem.* **1960,** *72,* 359–372.
251. Dimroth, O. *Justus Liebigs Ann. Chem.* **1910,** *373,* 336–370.
252. Dimroth, O. *Justus Liebigs Ann. Chem.* **1908,** *364,* 183–202.
253. Huisgen, R.; Sauer, J.; Sturm, H. J.; Markgraf, J. H. *Chem. Ber.* **1960,** *93,* 2106–2124.
254. *Review and Classification of 1,5-Electrocyclizations:* Huisgen, R. *Angew. Chem. Int. Ed. Engl.* **1980,** *19,* 947–973.
255. Sturm, H. J. Ph.D. Thesis, University of Munich, 1960.
256. Sauer, J.; Huisgen, R.; Sturm, H. J. *Tetrahedron* **1960,** *11,* 241–251.
257. Huisgen, R.; Axen, C.; Seidl, H. *Chem. Ber.* **1965,** *98,* 2966–2984.
258. Huisgen, R.; Sauer, J.; Seidel, M. *Chem. Ber.* **1960,** *93,* 2885–2891.
259. Huisgen, R.; Sturm, H. J.; Seidel, M. *Chem. Ber.* **1961,** *94,* 1555–1562.
260. Huisgen, R.; Seidel, M. *Chem. Ber.* **1961,** *94,* 2509–2511.
261. Huisgen, R.; Seidel, M.; Sauer, J.; McFarland, J. W.; Wallbillich, G. *J. Org. Chem.* **1959,** *24,* 892–893.
262. Huisgen, R.; Sauer, J.; Seidel, M. *Chem. Ber.* **1961,** *94,* 2503–2509.
263. Huisgen, R.; Seidel, M.; Wallbillich, G.; Knupfer, H. *Tetrahedron* **1962,** *17,* 3–29.
264. Clovis, J. S.; Eckell, A.; Huisgen, R.; Sustmann, R. *Chem. Ber.* **1967,** *100,* 60–70.
265. Huisgen, R.; Weberndörfer, V. *Chem. Ber.* **1967,** *100,* 71–78.
266. Meier, H.; Heinzelmann, W.; Heimgartner, H. *Chimia* **1980,** *34,* 504–506.
267. Toubro, N. H.; Holm, A. *J. Am. Chem. Soc.* **1980,** *102,* 2093–2094.
268. Wentrup, C.; Damerius, A.; Reichen, W. *J. Org. Chem.* **1978,** *43,* 2037–2041.
269. Sicard, G.; Baceiredo, A.; Bertrand, G. *J. Am. Chem. Soc.* **1988,** *110,* 2663.
270. Huisgen, R. *10 Jahre Fonds der Chemischen Industrie;* Düsseldorf, Germany, 1960; pp 73–102.
271. Huisgen, R. "Centenary Lecture, London, Dec. 8, 1960''; *Proc. Chem. Soc.* **1961,** 357–369.
272. Eckell, A.; Huisgen, R.; Sustmann, R.; Wallbillich, G.; Grashey, D.; Spindler, E. *Chem. Ber.* **1967,** *100,* 2192–2213.
273. Fliege, W.; Grashey, R.; Huisgen, R. *Chem. Ber.* **1984,** *117,* 1194–1214.
274. *Review:* Caramella, P.; Grünanger, P. In *1,3-Dipolar Cycloaddition Chemistry;* Padwa, A., Ed.; Wiley: New York, 1984; Vol. 1, pp 291–392.
275. Stollé, R.; Henke-Stark, F. *J. Prakt. Chem.* [2] **1930,** *124,* 261–300.
276. Wolff, L. *Justus Liebigs Ann. Chem.* **1912,** *394,* 59–68.
277. Huisgen, R.; Vossius, D.; Appl, M. *Angew. Chem.* **1955,** *67,* 756–757.
278. Huisgen, R.; Vossius, D.; Appl, M. *Chem. Ber.* **1958,** *91,* 1–12.
279. Doering, W. von E.; Odum, R. A. *Tetrahedron* **1966,** *22,* 81–93.
280. Huisgen, R.; Appl, M. *Chem. Ber.* **1958,** *91,* 12–21.
281. Appl, M.; Huisgen, R. *Chem. Ber.* **1959,** *92,* 2961–2967.

282. Smolinsky, G. *J. Org. Chem.* **1962**, *27*, 3557–3559.
283. Shillady, D. D.; Trindl, C. *Theor. Chim. Acta* **1976**, *43*, 137–144.
284. *Review:* Wentrup, C. *Adv. Heterocycl. Chem.* **1981**, *28*, 231–361.
285. Crow, W. D.; Wentrup, C. *Tetrahedron Lett.* **1968**, 6149–6152.
286. Chapman, O. L.; Le Roux, J. P. *J. Am. Chem. Soc.* **1978**, *100*, 282–285.
287. Michael, A. *J. Prakt. Chem.* [2] **1893**, *48*, 94–95.
288. Alder, K.; Stein, G. *Justus Liebigs Ann. Chem.* **1933**, *501*, 1–48.
289. Huisgen, R.; Möbius, L.; Müller, G.; Stangl, H.; Szeimies, G.; Vernon, J. M. *Chem. Ber.* **1965**, *98*, 3992–4013.
290. Fusco, R.; Bianchetti, G.; Pocar, D. *Gazz. Chim. Ital.* **1961**, *91*, 849–865, 933–957.
291. Fusco, R.; Bianchetti, G.; Pocar, D.; Ugo, R. *Chem. Ber.* **1963**, *96*, 802–812.
292. Huisgen, R.; Möbius, L.; Szeimies, G. *Chem. Ber.* **1965**, *98*, 1138–1152.
293. *Review:* L'abbé, G. *Ind. Chim. Belge* **1969**, *34*, 519–530.
294. Huisgen, R.; Szeimies, G. *Chem. Ber.* **1965**, *98*, 1153–1158.
295. Huisgen, R.; Szeimies, G.; Möbius, L. *Chem. Ber.* **1966**, *99*, 475–490.
296. Szeimies, G.; Huisgen, R. *Chem. Ber.* **1966**, *99*, 491–503.
297. Huisgen, R.; Szeimies, G.; Möbius, L. *Chem. Ber.* **1967**, *100*, 2494–2507.
298. Huisgen, R. "1,3-Dipolar Cycloadditions—Introduction, Survey, Mechanism"; In *1,3-Dipolar Cycloaddition Chemistry*; Padwa, A., Ed.; Wiley: New York, 1984; Vol. 1, pp 1–176. (a) 15–27; (b) 31–35; (c) 44–47; (d) 61–76; (e) 76–90; (f) 128–145; (g) 148–149; (h) 149–159.
299. Garfield, E. *Curr. Contents* **1974**, *10*, 5–12; **1977**, *51*, 5–20; **1986**, *45*, 3–11.
300. *Review:* Huisgen, R. *Angew. Chem. Int. Ed. Engl.* **1963**, *2*, 633–645.
301. Edward, J. T. *CHEMTECH* **1992**, 534–539.
302. Desimoni, G.; Tacconi, G.; Barco, A.; Pollini, G. P. *Natural Products Synthesis through Pericyclic Reactions*; ACS Monograph 180; American Chemical Society: Washington, DC, 1983; pp 181–187.
303. Mulzer, J. "Naturstoffe via 1,3-dipolare Cycloadditionen"; *Nachr. Chem. Tech.* **1984**, *32*, 882–887, 961–965.
304. Huisgen, R.; Stangl, H.; Sturm, H. J.; Wagenhofer, H. *Angew. Chem. Int. Ed. Engl.* **1962**, *1*, 50–51.
305. *Review:* Quilico, A. "Isoxazoles and Related Compounds"; In *The Chemistry of Heterocyclic Compounds*; Weissberger, A., Ed.; Interscience: New York, 1962; Vol. 17, pp 1–176.
306. Huisgen, R.; Mack, W. *Tetrahedron Lett.* **1961**, 583–586.
307. Huisgen, R.; Christl, M. (a) *Angew. Chem. Int. Ed. Engl.* **1967**, *6*, 456–457; (b) *Chem. Ber.* **1973**, *106*, 3291–3311.
308. Huisgen, R.; Grashey, R.; Steingruber, E. *Tetrahedron Lett.* **1963**, 1441–1445.
309. Huisgen, R.; Grashey, R.; Krischke, R. *Tetrahedron Lett.* **1962**, *3*, 387–391.
310. Huisgen, R. *J. Org. Chem.* **1968**, *33*, 2291–2297.
311. Hoffmann, R.; Woodward, R. B. *J. Am. Chem. Soc.* **1965**, *87*, 2046–2048.
312. Houk, K. N.; Yamaguchi, Z. In *1,3-Dipolar Cycloaddition Chemistry*; Padwa, A., Ed.; Wiley: New York, 1984; Vol. 2, pp 408–427.
313. *Review;* Hansen, H.-J.; Heimgartner, H. In *1,3-Dipolar Cycloaddition Chemistry*; Padwa, A., Ed.; Wiley: New York, 1984; Vol. 1, Chapter 2.
314. Criegee, R. *Justus Liebigs Ann. Chem.* **1953**, *583*, 1–36.
315. Criegee, R. *Angew. Chem. Int. Ed. Engl.* **1975**, *14*, 745–752.

316. *Review*: Meier, H.; Zeller, K.-P. *Angew. Chem. Int. Ed. Engl.* **1975**, *14*, 32–43.

317. Huisgen, R.; Binsch, G.; Ghosez, L. *Chem. Ber.* **1964**, *97*, 2628–2639.

318. Huisgen, R.; König, H.; Binsch, G.; Sturm, H. J. *Angew. Chem.* **1961**, *73*, 368–371.

319. Chaimovich, H.; Vaughan, R. J.; Westheimer, F. H. *J. Am. Chem. Soc.* **1968**, *90*, 4088–4093.

320. Huisgen, R.; Sturm, H. J.; Binsch, G. *Chem. Ber.* **1964**, *97*, 2864–2867.

321. Singh, B.; Ullman, E. F. *J. Am. Chem. Soc.* **1967**, *89*, 6911–6916.

322. Doering, W. von E.; Mole, T. *Tetrahedron* **1960**, *10*, 65–70.

323. Süs, O. *Justus Liebigs Ann. Chem.* **1944**, *556*, 65–84 and later papers.

324. Huisgen, R.; Binsch, G.; König, H. *Chem. Ber.* **1964**, *97*, 2868–2883, 2884–2892.

325. Binsch, G.; Huisgen, R.; König, H. *Chem. Ber.* **1964**, *97*, 2893–2902.

326. Lwowski, W.; Mattingly, T. W. *Tetrahedron Lett.* **1962**, 277–280.

327. Hafner, K.; König, C. *Angew. Chem. Int. Ed. Engl.* **1963**, *2*, 96.

328. Huisgen, R.; Blaschke, H. *Chem. Ber.* **1965**, *98*, 2985–2997.

329. Huisgen, R.; Blaschke, H. *Justus Liebigs Ann. Chem.* **1965**, *686*, 145–153.

330. *Review*: Ollis, W. D.; Ramsden, C. A. *Adv. Heterocycl. Chem.* **1976**, *19*, 1–122.

331. Huisgen, R.; Grashey, R.; Gotthardt, H.; Schmidt, R. *Angew. Chem. Int. Ed. Engl.* **1962**, *1*, 48–49.

332. Huisgen, R.; Gotthardt, H.; Grashey, R. *Angew. Chem. Int. Ed. Engl.* **1962**, *1*, 49.

333. Huisgen, R.; Gotthardt, H. *Chem. Ber.* **1968**, *101*, 1059–1071.

334. Huisgen, R.; Gotthardt, H.; Grashey, R. *Chem. Ber.* **1968**, *101*, 536–551.

335. Gotthardt, H.; Huisgen, R. *Chem. Ber.* **1968**, *101*, 552–563.

336. Huisgen, R.; Gotthardt, H. *Chem. Ber.* **1968**, *101*, 839–846.

337. Huisgen, R.; Gotthardt, H.; Bayer, H. O. *Angew. Chem. Int. Ed. Engl.* **1964**, *3*, 135–136.

338. Huisgen, R.; Gotthardt, H.; Bayer, H. O.; Schaefer, F. C. *Angew. Chem. Int. Ed. Engl.* **1964**, *3*, 136–137.

339. Bayer, H. O.; Huisgen, R.; Knorr, R.; Schaefer, F. C. *Chem. Ber.* **1970**, *103*, 2581–2597.

340. Gotthardt, H.; Huisgen, R.; Bayer, H. O. *J. Am. Chem. Soc.* **1970**, *92*, 4340–4344.

341. Huisgen, R.; Gotthardt, H.; Bayer, H. O. (a)*Tetrahedron Lett.* **1964**, 481–485; (b) *Chem. Ber.* **1970**, *103*, 2368–2387.

342. Huisgen, R.; Gotthardt, H.; Bayer, H. O.; Schaefer, F. C. *Chem. Ber.* **1970**, *103*, 2611–2624.

343. Gotthardt, H.; Huisgen, R. *Chem. Ber.* **1970**, *103*, 2625–2638.

344. Knorr, R.; Huisgen, R. *Chem. Ber.* **1970**, *103*, 2598–2610.

345. Funke, E.; Huisgen, R.; Schaefer, F. C. *Chem. Ber.* **1971**, *104*, 1550–1561.

346. Brunn, E.; Funke, E.; Gotthardt, H.; Huisgen, R. *Chem. Ber.* **1971**, *104*, 1562–1572.

347. Funke, E.; Huisgen, R. *Chem. Ber.* **1971**, *104*, 3222–3228.

348. Potts, K. T. In *1,3-Dipolar Cycloaddition Chemistry*; Padwa, A., Ed.; Wiley: New York, 1984; Vol. 2, pp 1–82.

349. Ollis, W. D.; Stanforth, S. P.; Ramsden, C. A. *Tetrahedron* **1985**, *41*, 2239–2329.

350. Grigg, R.; Gunaratne, H. Q. N.; Kemp, J. *J. Chem. Soc. Perkin Trans. 1* **1984**, 41–46 and later papers.

351. *Review:* Woodward, R. B.; Hoffmann, R. *Angew. Chem. Int. Ed. Engl.* **1969**, *8*, 781–853.

352. Huisgen, R. *J. Org. Chem.* **1977**, *41*, 403–419.

353. Sustmann, R.; Trill, H. *Angew. Chem. Int. Ed. Engl.* **1972**, *11*, 838–840.

354. Sustmann, R. *Pure Appl. Chem.* **1974**, *40*, 569–593.

355. Geittner, J.; Huisgen, R.; Sustmann, R. *Tetrahedron Lett.* **1977**, 881–884.

356. Bihlmaier, W.; Huisgen, R.; Reissig, H.-U.; Voss, S. *Tetrahedron Lett.* **1979**, 2621–2624.

357. Sustmann, R.; Wenning, E.; Huisgen, R. *Tetrahedron Lett.* **1977**, 877–880.

358. Bastide, J.; El Ghandour, N.; Henri-Rousseau, O. *Bull. Soc. Chim. Fr.* **1973**, 2290–2293.

359. Houk, K. N.; Sims, J.; Watts, C. R.; Luskus, L. J. *J. Am. Chem. Soc.* **1973**, *95*, 7301–7315.

360. *Review:* Sustmann, R.; Sicking, W. *Chem. Ber.* **1987**, *120*, 1653–1658.

361. Firestone, R. A. *J. Org. Chem.* **1968**, *33*, 2285–2290.

362. Firestone, R. A. *J. Org. Chem.* **1972**, *37*, 2181–2191.

363. Firestone, R. A. *Tetrahedron* **1977**, *33*, 3009–3039.

364. Bergmann, E.; Magat, M.; Wagenberg, D. *Ber. Dtsch. Chem. Ges.* **1930**, *63*, 2576–2584.

365. Schönberg, A.; Černik, D.; Urban, W. *Ber. Dtsch. Chem. Ges.* **1931**, *64*, 2577–2581.

366. Schönberg, A.; König, B.; Singer, E. *Chem. Ber.* **1967**, *100*, 767–777.

367. Kalwinsch. I.; Li, X.; Gottstein, J.; Huisgen, R. *J. Am. Chem. Soc.* **1981**, *103*, 7032–7033.

368. Li, X.; Huisgen, R. *Tetrahedron Lett.* **1983**, *24*, 4181–4184.

369. Huisgen, R.; Li, X. *Tetrahedron Lett.* **1983**, *24*, 4185–4188.

370. Huisgen, R.; Langhals, E. *Tetrahedron Lett.* **1989**, *30*, 5369–5372.

371. (a) Huisgen, R.; Seidl, H.; Brüning, I. *Chem. Ber.* **1969**, *102*, 1102–1116; (b) Huisgen, R.; Fisera, L.; Brüntrup, G., University of Munich, unpublished results, 1993.

372. (a) Sauer, J.; Schatz, J., University of Regensburg, unpublished results, 1993; (b) *Review:* Zwanenburg, B.; Lenz, B. G.; In *Methoden der Organischen Chemie (Houben-Weyl)*, 4th ed. G. Thieme: Stuttgart, Germany, 1985; Vol. E11, pp 911–949; (c) Huisgen, R.; Mloston, G.; Polborn, K., University of Munich, unpublished results, 1991.

373. Huisgen, R.; Rapp, J. *J. Am. Chem. Soc.* **1987**, *109*, 902–903.

374. Bihlmaier, W.; Geittner, J.; Huisgen, R.; Reissig, H.-U. *Heterocycles* **1978**, *10*, 147–152.

375. Dorn, H.; Ozegowski, R.; Gründemann, E. *J. Prakt. Chem.* **1979**, *321*, 555–564.

376. Huisgen, R.; Weinberger, R. *Tetrahedron Lett.* **1985**, *26*, 5119–5122.

377. Houk, K. N.; Firestone, R. A.; Munchausen, L. L.; Mueller, P. H.; Arison, B. H.; Garcia, L. A. *J. Am. Chem. Soc.* **1985**, *107*, 7227–7228.

378. Huisgen, R.; Mloston, G.; Langhals, E. *J. Am. Chem. Soc.* **1984**, *108*, 6401–6402.

379. Mloston, G.; Langhals, E.; Huisgen, R. *Tetrahedron Lett.* **1989**, *30*, 5373–5376.

380. Huisgen, R.; Mloston, G.; Langhals, E. *J. Org. Chem.* **1986**, *51*, 4085–4087.
381. (a) Huisgen, R.; Langhals, E.; Mloston, G.; Oshima, T. *Heterocycles* **1989**, *29*, 2069–2074; (b) Huisgen, R.; Langhals, E.; Nöth, H. *J. Org. Chem.* **1990**, *55*, 1412–1414; (c)Huisgen, R.; Langhals, E.; Oshima, T. *Heterocycles* **1989**, *29*, 2075–2078.
382. *Review:* Sauer, J.; Sustmann, R. *Angew. Chem. Int. Ed. Engl.* **1980**, *19*, 779–807.
383. *Review:* Huisgen, R. *Pure Appl. Chem.* **1981**, *53*, 171–187.
384. Huisgen, R.; Ooms, P. H. J.; Mingin, M.; Allinger, N. L. *J. Am. Chem. Soc.* **1980**, *102*, 3951–3953.
385. Fliege, W.; Huisgen, R. *Justus Liebigs Ann. Chem.* **1973**, 2038–2047.
386. Nuber, A. Ph.D. Thesis, University of Munich, 1983.
387. Schleyer, P. v. R. *J. Am. Chem. Soc.* **1967**, *89*, 701–703.
388. Inagaki, S.; Fujimoto, H.; Fukui, K. *J. Am. Chem. Soc.* **1976**, *98*, 4054–4061.
389. Wipff, G.; Morokuma, K. *Tetrahedron Lett.* **1980**, 4445–4448.
390. Spanget-Larsen, J.; Gleiter, R. *Tetrahedron Lett.* **1982**, 2435–2438.
391. Rondan, N. G.; Paddon-Row, M. N.; Caramella, P.; Mareda, J.; Mueller, P. H.; Houk, K. N. *J. Am. Chem. Soc.* **1982**, *104*, 4974–4976.
392. Watson, W. H.; Galloy, J.; Bartlett, P. D.; Roof, A. A. M. *J. Am. Chem. Soc.* **1981**, *103*, 2022–2031.
393. Rondan, N. G.; Paddon-Row, M. N.; Caramella, P.; Houk, K. N. *J. Am. Chem. Soc.* **1981**, *103*, 2436–2438.
394. Boeckh, D. Ph.D. Thesis, University of Munich, 1986.
395. Leitich, J. *Angew. Chem. Int. Ed. Engl.* **1976**, *15*, 372–373.
396. Huisgen, R.; Palacios Gambra, F. *Chem. Ber.* **1982**, *115*, 2242–2255.
397. Whitesides, G. M.; Goe, G. L.; Cope, A. C. *J. Am. Chem. Soc.* **1969**, *91*, 2608–2616.
398. (a) Huisgen, R.; Boeckh, D.; Nöth, H. *J. Am. Chem. Soc.* **1987**, *109*, 1248–1249; (b) Benson, S. W.; Cruickshank, F. R.; Golden, D. M.; Haugen, G. R.; O'Neal, H. E.; Rodgers, A. S.; Shaw, R.; Walsh, R. *Chem. Rev.* **1969**, *69*, 279–374.
399. Dewar, M. J. S. *Angew. Chem. Int. Ed. Engl.* **1971**, *10*, 761–776.
400. Zimmerman, H. E. *Acc. Chem. Res.* **1971**, *4*, 272–280.
401. Fukui, K. *Bull. Chem. Soc. Jpn.* **1966**, *39*, 498–503.
402. Woodward, R. B.; Hoffmann, R. *J. Am. Chem. Soc.* **1965**, *87*, 395–397.
403. Willstätter, R.; Waser, E. *Ber. Dtsch. Chem. Ges.* **1911**, *44*, 3423–3445.
404. Reppe, W.; Schlichting, O.; Klager, K.; Toepel, T. *Justus Liebigs Ann. Chem.* **1948**, *560*, 1–92.
405. Huisgen, R.; Mietzsch, F. *Angew. Chem. Int. Ed. Engl.* **1964**, *3*, 83–85.
406. Karle, I. L. *J. Chem. Phys.* **1952**, *20*, 65–70.
407. Vogel, E.; Kiefer, H.; Roth, W. R. *Angew. Chem. Int. Ed. Engl.* **1964**, *3*, 442–443.
408. *Review:* Huisgen, R.; Mietzsch, F.; Boche, G.; Seidl, H. In *Organic Reaction Mechanisms;* Chemical Society of London: London, 1965; Spec. Publ. No. 19, pp 3–20.
409. Squillacote, M. E.; Bergman, A. *J. Org. Chem.* **1986**, *51*, 3911–3913.
410. *Review:* Maier, G. *Valenzisomerisierungen;* Verlag Chemie: Weinheim, Germany, 1972; pp 105–117.

411. England, D. C. University of Munich, unpublished results, 1962.
412. Hentschel, W. Ph.D. Thesis, University of Munich, 1973.
413. Rubin, M. B. *J. Am. Chem. Soc.* **1981,** *103,* 7791–7792.
414. Huisgen, R.; Juppe, G. *Chem. Ber.* **1961,** *94,* 2332–2349.
415. Cope, A. C.; Haven, A. C.; Ramp, F. L.; Trumbull, E. R. *J. Am. Chem. Soc.* **1952,** *74,* 4867–4871.
416. Huisgen, R.; Boche, G.; Dahmen, A.; Hechtl, W. *Tetrahedron Lett.* **1968,** 5215–5219.
417. Cope, A. C.; Burg, M. *J. Am. Chem. Soc.* **1952,** *74,* 168–172.
418. Huisgen. R.; Konz, W. E. *J. Am. Chem. Soc.* **1970,** *92,* 4102–4104.
419. Konz, W. E.; Hechtl, W.; Huisgen, R. *J. Am. Chem. Soc.* **1970,** *92,* 4104–4105.
420. Huisgen. R.; Konz, W. E.; Gream, G. E. *J. Am. Chem. Soc.* **1970,** *92,* 4105–4106.
421. Huisgen, R.; Boche, G.; Hechtl, W.; Huber, H. *Angew. Chem. Int. Ed. Engl.* **1966,** *5,* 585–586.
422. Huisgen, R.; Boche, G. *Tetrahedron Lett.* **1965,** 1769–1774.
423. Gasteiger, J.; Gream, G. E.; Huisgen, R.; Konz, W. E.; Schnegg, U. *Chem. Ber.* **1971,** *104,* 2412–2419.
424. *Review:* Winstein, S. In *Aromaticity;* Chemical Society of London: London, 1967; Spec. Publ. No. 21, pp 5–45.
425. Winstein, S.; Kreiter, C. G.; Brauman, J. I. *J. Am. Chem. Soc.* **1966,** *88,* 2047–2048.
426. Boche, G.; Hechtl, W.; Huisgen, R. *J. Am. Chem. Soc.* **1967,** *89,* 3344–3345.
427. Huisgen, R.; Boche, G.; Huber, H. *J. Am. Chem. Soc.* **1967,** *89,* 3345–3346.
428. Huisgen, R.; Gasteiger, J. *Angew. Chem. Int. Ed. Engl.* **1972,** *11,* 1104–1105.
429. Olah, G. A.; Staral, J. S.; Liang, G.; Paquette, L. A.; Melega, W. P.; Carmody, M. J. *J. Am. Chem. Soc.* **1977,** *99,* 3349–3355.
430. Huisgen, R.; Konz, W. E.; Schnegg, U. *Angew. Chem. Int. Ed. Engl.* **1972,** *11,* 715–716.
431. Gasteiger, J.; Huisgen, R. *Angew. Chem. Int. Ed. Engl.* **1972,** *11,* 716–717.
432. Gasteiger, J.; Huisgen, R. *J. Am. Chem. Soc.* **1972,** *94,* 6541–6543.
433. Schnegg, U. Ph.D. Thesis, University of Munich, 1974.
434. Jensen, F. R.; Coleman, W. E. *J. Am. Chem. Soc.* **1958,** *80,* 6149.
435. Huisgen, R.; Seidl, H. *Tetrahedron Lett.* **1964,** 3381–3386.
436. Vogel, E. *Justus Liebigs Ann. Chem.* **1958,** *615,* 14–21.
437. Criegee, R.; Noll, K. *Justus Liebigs Ann. Chem.* **1959,** *627,* 1–14.
438. Quinkert, G.; Finke, M.; Palmowski, J.; Wiersdorff, W.-W. *Mol. Photochem.* **1969,** *1,* 433–460.
439. Takahashi, Y.; Kochi, J. K. *Chem. Ber.* **1988,** *121,* 253–269.
440. Flynn, C. R.; Michl, J. *J. Am. Chem. Soc.* **1974,** *96,* 3280–3288.
441. Huisgen, R.; Dahmen, A.; Huber, H. *J. Am. Chem. Soc.* **1967,** *89,* 7130–7131.
442. Marvell, E. N.; Caple, G.; Schatz, B. *Tetrahedron Lett.* **1965,** 385–389.
443. Huisgen, R.; Dahmen, A.; Huber, H. *Tetrahedron Lett.* **1969,** 1461–1464.
444. Dahmen, A.; Huisgen, R. *Tetrahedron Lett.* **1969,** 1465–1469.
445. Bandaranayake, W. M., Banfield, J. E.; Black, D. St. C. *J. Chem. Soc. Chem. Commun.* **1980,** 902–903.
446. Nicolaou, K. C.; Patasis, N. C.; Zipkin, R. E. *J. Am. Chem. Soc.* **1982,** *104,* 5560–5562.

447. Bannwarth, W.; Eidenschink, R.; Kauffmann, T. *Angew. Chem. Int. Ed. Engl.* **1974,** *13,* 468–469.

448. Boche, G.; Buckl, K.; Martens, D.; Schneider, D. R.; Wagner, H.-U. *Chem. Ber.* **1979,** *112,* 2961–2996.

449. Heine, H. W.; Peavy, R. E. *Tetrahedron Lett.* **1965,** 3123–3126.

450. Padwa, A.; Hamilton, L. *Tetrahedron Lett.* **1965,** 4363–4367.

451. Huisgen, R.; Scheer, W.; Szeimies, G.; Huber, H. *Tetrahedron Lett.* **1966,** 397–404.

452. Huisgen, R.; Scheer, W.; Huber, H. *J. Am. Chem. Soc.* **1967,** *89,* 1753–1755.

453. Huisgen, R.; Mäder, H. *Angew. Chem. Int. Ed. Engl.* **1969,** *8,* 604–606.

454. Huisgen, R.; Scheer, W.; Mäder, H. *Angew. Chem. Int. Ed. Engl.* **1969,** *8,* 602–604.

455. Huisgen, R.; Mäder, H. *J. Am. Chem. Soc.* **1971,** *93,* 1777–1779.

456. Hermann, H.; Huisgen, R.; Mäder, H. *J. Am. Chem. Soc.* **1971,** *93,* 1779–1780.

457. *Review:* Huisgen, R. *XIIIth International Congress of Pure and Applied Chemistry;* Butterworths: London, 1971; Vol. 1, pp 175–195.

458. Huisgen, R.; Scheer, W. *Tetrahedron Lett.* **1971,** 481–484.

459. Ross, C. H. Ph.D. Thesis, University of Munich, 1975.

460. Hall, J. H.; Huisgen, R.; Ross, C. H.; Scheer, W. *J. Chem. Soc. Chem. Commun.* **1971,** 1188–1190.

461. Seidl, H.; Huisgen, R.; Knorr, R. *Chem. Ber.* **1969,** *102,* 909–914.

462. Huisgen, R.; Niklas, K. *Heterocycles* **1984,** *22,* 21–26.

463. DoMinh, T.; Trozzolo, A. M. *J. Am. Chem. Soc.* **1972,** *94,* 4046–4048.

464. Trozzolo, A. M.; Leslie, T. M.; Sarpotdar, A. S.; Small, R. D.; Ferrandi, G. J.; DoMinh, T.; Hartless, R. L. *Pure Appl. Chem.* **1979,** *51,* 261–270.

465. *Review:* Lown, J. W. In *1,3-Dipolar Cycloaddition Chemistry;* Padwa, A., Ed.; Wiley: New York, 1984; Vol. 2, Chapter 6.

466. Schaap, A. P.; Prasad, G.; Siddiqui, S. *Tetrahedron Lett.* **1984,** *25,* 3035–3038.

467. Caër, V.; Laurent, A.; Laurent, E.; Tardivel, R.; Cebulska, Z.; Bartnik, R. *Nouv. J. Chim.* **1987,** *11,* 351–356.

468. Linn, W. J.; Benson, R. E. *J. Am. Chem. Soc.* **1965,** *87,* 3657–3665.

469. Linn, W. J. *J. Am. Chem. Soc.* **1965,** *87,* 3665–3672.

470. MacDonald, H. H. J.; Crawford, R. J. *Can. J. Chem.* **1972,** *50,* 428–433.

471. Hehre, W. J., University of California at Irvine, personal communication, 1971.

472. Hamberger, H.; Huisgen, R. *J. Chem. Soc. Chem. Commun.* **1971,** 1190–1192.

473. Ullman, E. F.; Milks, J. E. *J. Am. Chem. Soc.* **1964,** *86,* 3814–3819.

474. Dahmen, A.; Hamberger, H.; Huisgen, R.; Markowski, V. *J. Chem. Soc. Chem. Commun.* **1971,** 1192–1194.

475. Markowski, V.; Huisgen, R. *J. Chem. Soc. Chem. Commun.* **1977,** 439–440.

476. Huisgen, R.; Markowski, V. *J. Chem. Soc. Chem. Commun.* **1977,** 440–442.

477. *Review:* Huisgen, R. *Angew. Chem. Int. Ed. Engl.* **1977,** *16,* 572–585.

478. Huisgen, R.; Markowski, V.; Hermann, R. *Heterocycles* **1977,** *7,* 61–66.

479. Markowski, V.; Huisgen, R. *Tetrahedron Lett.* **1976,** 4643–4646.

480. Paladini, J. C.; Chuche, J. *Bull. Soc. Chim. Fr.* **1974,** 197–202.

481. DoMinh, T.; Trozzolo, A. M.; Griffin, G. W. *J. Am. Chem. Soc.* **1970,** *92,* 1402–1403.

482. Lee, G. A. *J. Org. Chem.* **1976**, *41*, 2656–2658.
483. Wong, J. P. K.; Fahmi, A. A.; Griffin, G. W.; Bhacca, N. S. *Tetrahedron* **1981**, *37*, 3345–3355.
484. Albini, A.; Arnold, D. R. *Can. J. Chem.* **1978**, *56*, 2985–2993.
485. Schaap, A. P.; Siddiqui, S.; Gagnon, S. D.; Lopez, L. *J. Am. Chem. Soc.* **1983**, *105*, 5149–5150.
486. Futamura, S.; Kusunose, S.; Ohta, H.; Kamiya, Y. *J. Chem. Soc. Perkin Trans. 1* **1984**, 15–19.
487. Buchner, E.; Curtius, Th. *Ber. Dtsch. Chem. Ges.* **1885**, *18*, 2371–2377.
488. Dieckmann, W. *Ber. Dtsch. Chem. Ges.* **1910**, *43*, 1024–1031.
489. de March, P.; Huisgen, R. *J. Am. Chem. Soc.* **1982**, *104*, 4952.
490. Huisgen, R.; de March, P. *J. Am. Chem. Soc.* **1982**, *104*, 4953–4954.
491. Bronberger, F. Ph.D. Thesis, University of Munich, 1985.
492. Mitra, A., University of Munich, experiments 1984–1985.
493. Janulis, E. P.; Arduengo, A. J., III *J. Am. Chem. Soc.* **1983**, *105*, 5929–5930.
494. Hamaguchi, M.; Ibata, T. *Tetrahedron Lett.* **1974**, 4475–4476 and later papers.
495. Laurent, M. A. *C. R. Seances Acad. Sci.* **1844**, *19*, 353.
496. Borodin, A. *Ann. Chem. Pharm.* **1859**, *110*, 78–85.
497. Hunter, D. A.; Sim, S. K.; Steiner, R. P. *Can. J. Chem.* **1977**, *55*, 1229–1241.
498. Reimlinger, H. *Chem. Ber.* **1970**, *103*, 1900–1907.
499. Taylor, E. C.; Turchi, I. T. *Chem. Rev.* **1979**, 181–231.
500. Marvell, E. N. *Thermal Electrocyclic Reactions;* Academic: Orlando, FL, 1980; Chapter 10.
501. Thiele, J. *Justus Liebigs Ann. Chem.* **1892**, *270*, 1–63.
502. Butler, R. N. *Adv. Heterocycl. Chem.* **1976**, *20*, 402–424.
503. Doering, W. von E.; Roth, W. R. *Tetrahedron* **1962**, *18*, 67–74.
504. *Review:* Sauer, J. *Angew. Chem. Int. Ed. Engl.* **1967**, *6*, 16–33.
505. *Review:* Bartlett, P. D. *Science* **1968**, *159*, 833–838.
506. Bartlett, P. D. *Quart. Rev. Chem. Soc.* **1970**, *24*, 473–497.
507. *Review:* Huisgen, R. In *Topics in Heterocyclic Chemistry;* Castle, R. N., Ed.; Wiley: New York, 1969; pp 223–252.
508. Diels, O.; Alder, K. *Justus Liebigs Ann. Chem.* **1932**, *498*, 16–49.
509. Diels, O.; Harms, J. *Justus Liebigs Ann. Chem.* **1936**, *525*, 73–94.
510. Acheson, R. M., Plunkett, A. O. *J. Chem. Soc.* **1964**, 2676–2683.
511. Huisgen, R.; Morikawa, M.; Herbig, K.; Brunn, E. *Chem. Ber.* **1967**, *100*, 1094–1106.
512. Huisgen, R.; Herbig, K.; Morikawa, M. *Chem. Ber.* **1967**, *100*, 1107–1115.
513. Huisgen, R.; Herbig, K. *Justus Liebigs Ann. Chem.* **1965**, *688*, 98–112.
514. Huisgen, R.; Morikawa, M.; Breslow, D. S.; Grashey, R. *Chem. Ber.* **1967**, *100*, 1602–1615.
515. Morikawa, M.; Huisgen, R. *Chem. Ber.* **1967**, *100*, 1616–1620.
516. Breslow, D. S., University of Munich, unpublished results, 1965.
517. Regnault, V. *Ann. de Chim.* [2] **1837**, *65*, 98–112.
518. Bordwell, F. G.; Colton, F. B.; Knell, M. *J. Am. Chem. Soc.* **1954**, *76*, 3950–3952.
519. Eitner, P. *Ber. Dtsch. Chem. Ges.* **1892**, *25*, 461–472.
520. Gotthardt, H.; Schenk, K. H. *Chem. Ber.* **1985**, *118*, 4567–4577.

521. Staudinger, H.; Suter, E. *Ber. Dtsch. Chem. Ges.* **1920**, *53*, 1092–1105.
522. Huisgen, R.; Feiler, L. A. *Chem. Ber.* **1969**, *102*, 3391–3404.
523. Huisgen, R.; Feiler, L. A.; Otto, P. *Chem. Ber.* **1969**, *102*, 3405–3427.
524. Feiler, L. A.; Huisgen, R.; Koppitz, P. *J. Chem. Soc. Chem. Commun.* **1974**, 405–406.
525. Mayr, H.; Huisgen, R. *Angew. Chem. Int. Ed. Engl.* **1975**, *14*, 499–500.
526. Huisgen, R.; Mayr, H. *J. Chem. Soc. Chem. Commun.* **1976**, 55–56.
527. Mayr, H.; Huisgen, R. *J. Chem. Soc. Chem. Commun.* **1976**, 57–58.
528. Huisgen, R.; Feiler, L. A.; Binsch, G. *Angew. Chem. Int. Ed. Engl.* **1964**, *3*, 753–754.
529. Huisgen, R.; Feiler, L. A.; Binsch, G. *Chem. Ber.* **1969**, *102*, 3460–3474.
530. Montaigne, R.; Ghosez, L. *Angew. Chem. Int. Ed. Engl.* **1968**, *7*, 221.
531. Sustmann, R.; Ansmann, A.; Vahrenholt, F. *J. Am. Chem. Soc.* **1972**, *94*, 8099–8105.
532. Woodward, R. B. In *Aromaticity* (Sheffield Symposium); Ollis, W. D., Ed.; Chemical Society: Sheffield, England, 1967; Chem. Soc. Spec. Publ. 21, pp 217–249.
533. Huisgen, R.; Feiler, L. A.; Otto, P. *Chem. Ber.* **1969**, *102*, 3444–3459.
534. Effenberger, F.; Prossel, G.; Fischer, P. *Chem. Ber.* **1971**, *104*, 2002–2012.
535. Huisgen, R.; Mayr, H. *Tetrahedron Lett.* **1975**, 2965–2968.
536. Mayr, R.; Huisgen, R. *Tetrahedron Lett.* **1975**, 1349–1352.
     R. Mayr or H. Mayr?
537. Huisgen, R.; Mayr, H. *Tetrahedron Lett.* **1975**, 2969–2972.
538. Ghosez, L.; Dumont, W., 1972; quoted in reference 540.
539. Huisgen, R.; Otto, P. *Chem. Ber.* **1969**, *102*, 3475–3485.
540. *Review:* Ghosez, L.; O'Donnell, M. J. In *Pericyclic Reactions;* Marchand, A. P.; Lehr, R. E., Eds.; Academic Press: Orlando, FL, 1977; pp 79–140.
541. Stevens, H. C.; Reich, D. A.; Brandt, D. R.; Fountain, K. R.; Gaughan, E. J. *J. Am. Chem. Soc.* **1965**, *87*, 5257–5259.
542. Cossement, E.; Biname, R.; Ghosez, L. *Tetrahedron Lett.* **1974**, 997–1000.
543. Goldstein, S.; Vannes, P.; Houge, C.; Frisque-Hesbain, A. H.; Wiaux-Zamar, C.; Ghosez, L. *J. Am. Chem. Soc.* **1981**, *103*, 4616–4618.
544. Hasek, R. H.; Gott, P. G.; Martin, J. C. *J. Org. Chem.* **1964**, *29*, 2513–2516.
545. Otto, P.; Feiler, L. A.; Huisgen, R. *Angew. Chem. Int. Ed. Engl.* **1968**, *7*, 737–738.
546. Huisgen, R.; Otto, P. *J. Am. Chem. Soc.* **1969**, *91*, 5922–5923.
547. *Review:* Huisgen, R. *Pure Appl. Chem.* **1980**, *52*, 2283–2302.
548. England, D. C.; Krespan, C. G. *J. Org. Chem.* **1970**, *35*, 3312–3322.
549. Huisgen, R.; Otto, P. *J. Am. Chem. Soc.* **1968**, *90*, 5342–5343.
550. Staudinger, H. *Justus Liebigs Ann. Chem.* **1907**, *356*, 51–123.
551. Staudinger, H.; Klever, H. W.; Kober, P. *Justus Liebigs Ann. Chem.* **1910**, *374*, 1–39.
552. Martin, J. C.; Hoyle, V. A.; Brannock, K. C. *Tetrahedron Lett.* **1965**, 3589–3594.
553. Huisgen, R.; Davis, B. A.; Morikawa, M. *Angew. Chem. Int. Ed. Engl.* **1968**, *7*, 826–827.
554. Kristian, P., University of Munich, unpublished results, 1969.
555. *Review:* Huisgen, R. *Acc. Chem. Res.* **1977**, *10*, 117–124.
556. Salem, L.; Rowland, C. *Angew. Chem. Int. Ed. Engl.* **1972**, *11*, 92–112.

557. Segal, G. A. *J. Am. Chem. Soc.* **1974**, *96*, 7892–7898.
558. Hoffmann, R.; Swaminathan, S.; Odell, B. G.; Gleiter, R. *J. Am. Chem. Soc.* **1970**, *92*, 7091–7097.
559. Borden, W. T.; Davidson, E. R. *J. Am. Chem. Soc.* **1980**, *102*, 5409–5410.
560. Stephenson, L. M.; Gibson, T. A. *J. Am. Chem. Soc.* **1972**, *94*, 4599–4602.
561. Middleton, W. J. *J. Org. Chem.* **1965**, *30*, 1402–1407.
562. Proskow, S.; Simmons, H. E.; Cairns, T. L. *J. Am. Chem. Soc.* **1966**, *88*, 5254–5266.
563. Williams, J. K.; Wiley, D. W.; McKusick, B. C. *J. Am. Chem. Soc.* **1962**, *84*, 2210–2215.
564. Huisgen, R.; Steiner, G. *J. Am. Chem. Soc.* **1973**, *95*, 5054–5055.
565. Huisgen, R.; Graf, H. *J. Org. Chem.* **1979**, *44*, 2595–2596.
566. Pocker, Y.; Ellsworth, D. L. *J. Am. Chem. Soc.* **1977**, *99*, 2284–2293.
567. Huisgen, R.; Steiner, G. *J. Am. Chem. Soc.* **1973**, *95*, 5055–5056.
568. Huisgen, R.; Steiner, G. *Tetrahedron Lett.* **1973**, 3763–3768.
569. Stewart, C. A. *J. Org. Chem.* **1963**, *28*, 3320–3323.
570. Huisgen, R.; Plumet Ortega, J. *Tetrahedron Lett.* **1978**, 3975–3978.
571. Graf, H.; Huisgen, R. *J. Org. Chem.* **1979**, *44*, 2594–2595.
572. Schleyer, P. von R. *Pure Appl. Chem.* **1987**, *59*, 1647–1660.
573. Apeloig, Y.; Karni, M. *J. Chem. Soc. Perkin Trans. II* **1988**, 625–636.
574. Brückner, R.; Huisgen, R. *Tetrahedron Lett.* **1990**, *31*, 2557–2560, 2561–2564.
575. Brückner, R. Ph.D. Thesis, University of Munich, 1984.
576. Brückner, R.; Huisgen, R. *J. Org. Chem.* **1991**, *56*, 1677–1679.
577. Sauer, J.; Wiest, H.; Mielert, A. *Chem. Ber.* **1964**, *97*, 3183–3207.
578. Houk, K. N. *Acc. Chem. Res.* **1975**, *8*, 361–369.
579. Huisgen, R.; Schug, R. *J. Am. Chem. Soc.* **1976**, *98*, 7819–7821.
580. Steiner, G.; Huisgen, R. *J. Am. Chem. Soc.* **1973**, *95*, 5056–5058.
581. Reichardt, C. *Solvent Effects in Organic Chemistry*; Verlag Chemie: Weinheim, Germany, 1979.
582. Laidler, K. J.; Eyring, H. *Ann. N. Y. Acad. Sci.* **1940**, *39*, 303.
583. Huisgen, R.; Schug, R.; Steiner, G. *Angew. Chem. Int. Ed. Engl.* **1974**, *13*, 80–81.
584. Karle, I.; Flippen, J.; Huisgen, R.; Schug, R. *J. Am. Chem. Soc.* **1975**, *97*, 5285–5287.
585. Huisgen, R.; Schug, R.; Steiner, G. *Angew. Chem. Int. Ed. Engl.* **1974**, *13*, 81–82.
586. *Review:* Huisgen, R. *Acc. Chem. Res.* **1977**, *10*, 199–206.
587. Graf, H. Ph.D. Thesis, University of Munich, 1980, pp 134–157, 246–272.
588. Schug, R.; Huisgen, R. *J. Chem. Soc. Chem. Commun.* **1975**, 60–61.
589. Hall, H. K. *Angew. Chem. Int. Ed. Engl.* **1983**, *22*, 440–456.
590. Hall, H. K., Jr.; Padias, A. B. *Acc. Chem. Res.* **1990**, *23*, 3–9.
591. Huisgen, R.; Penelle, J.; Mloston, G.; Padias, A. B.; Hall, H. K., Jr. *J. Am. Chem. Soc.* **1992**, *114*, 266–274.
592. Oshima, T., University of Munich, unpublished experiments 1984–1985.
593. Urrutia Desmaison, G. Ph.D. Thesis, University of Munich, 1985.
594. Nöth, H., University of Munich, unpublished results.
595. Prelog, V.; Helmchen, G. *Angew. Chem. Int. Ed. Engl.* **1982**, *21*, 567–584.
596. Huisgen, R.; Mloston, G., University of Munich, unpublished results, 1990.

597. Ashok, K.; Huisgen, R.; Urrutia Desmaison, G., University of Munich, unpublished results, 1985, 1990.
598. Prantl, G.; Eibler, E.; Sauer, J. *Tetrahedron Lett.* **1982**, *23*, 1139–1142.
599. Brückner, R.; Huisgen, R. *Tetrahedron Lett.* **1991**, *32*, 1871–1874, 1875–1878.
600. Huisgen, R.; Brückner, R. *J. Org. Chem.* **1991**, *56*, 1679–1681.
601. Wieland, H.; Horner, L. *Justus Liebigs Ann. Chem.* **1937**, *528*, 73–100.
602. Robinson, R.; Stephen, A. M. *Nature London* **1948**, *162*, 177.
603. Woodward, R. B.; Cava, M. P.; Ollis, W. D.; Hunger, A.; Daeniker, H. U.; Schenker, K. *Tetrahedron* **1963**, *19*, 247–288.
604. Huisgen, R. *Angew. Chem.* **1959**, *71*, 5–6.
605. Witkop, B. *Justus Liebigs Ann. Chem.* **1992**, 1-32.
606. Huisgen, R. "Rudolf Criegee"; (a) *Jahrbuch Bayer. Akad. Wissenschaften* **1976**, 235–238; (b) *Chemie in unserer Zeit* **1978**, *2*, 49–55; (c) *J. Chem. Educ.* **1979**, *56*, 369–374.
607. Hoffmann, R. (a) *Am. Sci.* **1988**, *76*, 389, 604; **1989**, *77*, 177, 330; (b) *J. Aesthetics Art Crit.* **1990**, *48*:3, 191–204.
608. Hofacker, U. *Nachr. Chem. Tech. Lab.* (a) **1991**, *39*, 1422–1426; (b) **1992**, *40*, 415; (c) **1993**, 876.
609. Ochiai, E. *Aromatic Amine Oxides*; Elsevier: Amsterdam, Netherlands, 1967.
610. Huisgen, R. "Eiji Ochiai, 1898–1974" (Obituary); *Jahrbuch Bayer. Akad. Wissenschaften* **1976**, 215–217.
611. Huisgen, R. "David Ginsburg, 1920–1988" (Obituary); *Jahrbuch Bayer. Akad. Wissenschaften* **1988**, 218–221.
612. Huisgen, R. *Nachr. Chem. Tech. Lab.* **1989**. *37*, 1257.

Index

Copy editing: Colleen P. Stamm
Production: Margaret J. Brown
Indexing: A. L. McClellan

Production Manager: Cheryl R. Wurzbacher

Printed and bound by Maple Press, York, PA

Bestsellers from ACS Books

The ACS Style Guide: A Manual for Authors and Editors
Edited by Janet S. Dodd
264 pp; clothbound ISBN 0–8412–0917–0; paperback ISBN 0–8412–0943–X

The Basics of Technical Communicating
By B. Edward Cain
ACS Professional Reference Book; 198 pp;
clothbound ISBN 0–8412–1451–4; paperback ISBN 0–8412–1452–2

Chemical Activities (student and teacher editions)
By Christie L. Borgford and Lee R. Summerlin
330 pp; spiralbound ISBN 0–8412–1417–4; teacher ed. ISBN 0–8412–1416–6

Chemical Demonstrations: A Sourcebook for Teachers,
Volumes 1 and 2, Second Edition
Volume 1 by Lee R. Summerlin and James L. Ealy, Jr.;
Vol. 1, 198 pp; spiralbound ISBN 0–8412–1481–6;
Volume 2 by Lee R. Summerlin, Christie L. Borgford, and Julie B. Ealy
Vol. 2, 234 pp; spiralbound ISBN 0–8412–1535–9

Chemistry and Crime: From Sherlock Holmes to Today's Courtroom
Edited by Samuel M. Gerber
135 pp; clothbound ISBN 0–8412–0784–4; paperback ISBN 0–8412–0785–2

Writing the Laboratory Notebook
By Howard M. Kanare
145 pp; clothbound ISBN 0–8412–0906–5; paperback ISBN 0–8412–0933–2

Developing a Chemical Hygiene Plan
By Jay A. Young, Warren K. Kingsley, and George H. Wahl, Jr.
paperback ISBN 0–8412–1876–5

Introduction to Microwave Sample Preparation: Theory and Practice
Edited by H. M. Kingston and Lois B. Jassie
263 pp; clothbound ISBN 0–8412–1450–6

Principles of Environmental Sampling
Edited by Lawrence H. Keith
ACS Professional Reference Book; 458 pp;
clothbound ISBN 0–8412–1173–6; paperback ISBN 0–8412–1437–9

Biotechnology and Materials Science: Chemistry for the Future
Edited by Mary L. Good (Jacqueline K. Barton, Associate Editor)
135 pp; clothbound ISBN 0–8412–1472–7; paperback ISBN 0–8412–1473–5

For further information and a free catalog of ACS books, contact:
American Chemical Society
Distribution Office, Department 225
1155 16th Street, NW, Washington, DC 20036
Telephone 800–227–5558